Winter Key to Deciduous Woody Plants of East Texas

BY

MICHAEL S. FOUNTAIN

&

LANCE C. JONES

ILLUSTRATIONS BY

BRUCE LYNDON CUNNINGHAM

ARTHUR TEMPLE COLLEGE OF
FORESTRY & AGRICULTURE
STEPHEN F. AUSTIN STATE UNIVERSITY
NACOGDOCHES, TEXAS 75962-6109

A BRUCE LYNDON CUNNINGHAM PRODUCTION

www.forester-artist.com

Winter Key to the Deciduous Woody Plants of East Texas
First edition, October 2008

Library of Congress Control Number: 2008906054

ISBN-13:978-0-934115-04-9
ISBN: 0-934115-04-4

Bruce Lyndon Cunningham Productions
Forester-Artist.Com
180 County Road 8201
Nacogdoches, TX 75964
936.569.6965
The United States of America
www.forester-artist.com
brucelc@suddenlink.net

Printed in The United States of America

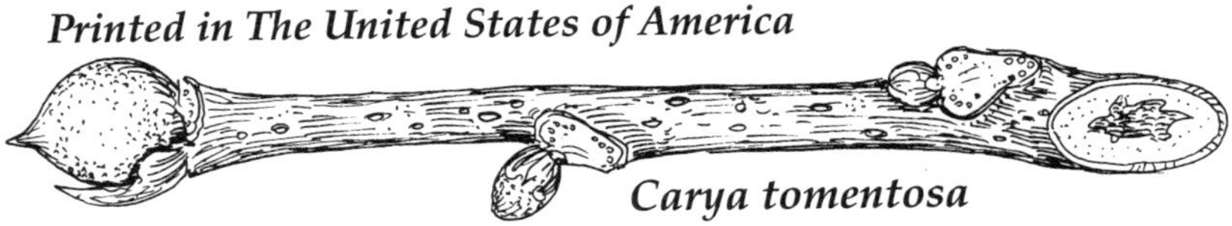

Carya tomentosa

Preface

This key began as a project for a special problems class when I was an undergraduate student at Stephen F. Austin State University. It then evolved when I became a graduate teaching assistant under Dr. Ray Hicks to become part of a plant identification manual that was utilized for many years as the textbook for dendrology labs at SFA but it is no longer in print and there is no other plant manual that contains a winter dichotomous key to the deciduous woody plants of east Texas. This key covers approximately 120+ species, the exact number is not available due to some species being keyed only to the genus level of classification. I am sure some will argue that there are other species that should have been included. All species that maintain persistent green foliage throughout the winter were intentionally omitted and some that are tardily deciduous were included while others were omitted. This decision was based on the condition most commonly encountered during the mid to late winter period. If you need a reference key for woody plants of east Texas in the growing season, I recommend *"Trees, Shrubs, and Woody Vines of East Texas"* by Elray S. Nixon and illustrated and published by Bruce Lyndon Cunningham (ISBN: 978-0-934115-01-8).

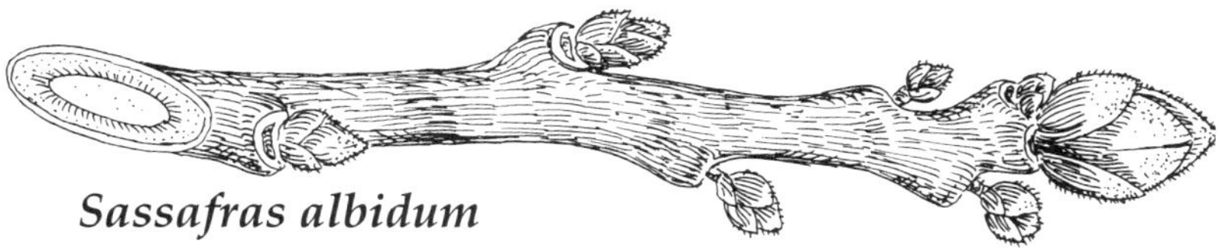

Sassafras albidum

Key to Front Cover

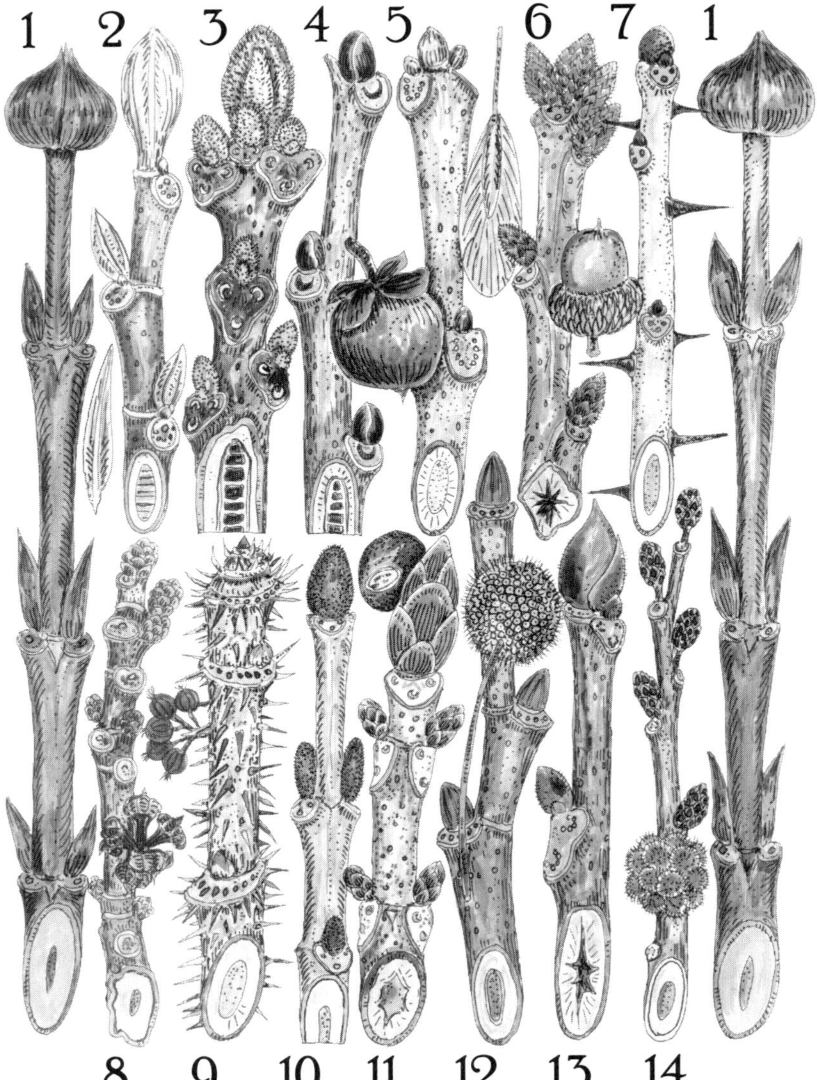

1. ***Cornus florida*** - flowering dogwood
2. ***Liriodendron tulipifera*** - yellow-poplar
3. ***Juglans nigra*** - black walnut
4. ***Diospyros virginiana*** - common persimmon
5. ***Fraxinus caroliniana*** - Carolina ash
6. ***Quercus michauxii*** - swamp chestnut oak
7. ***Zanthoxylum clava-herculis*** - Hercules-club
8. ***Taxodium distichum*** - baldcypress
9. ***Aralia spinosa*** - devil's walkingstick
10. ***Viburnum rufidulum*** - rusty blackhaw
11. ***Aesculus pavia*** - red buckeye
12. ***Platanus occidentalis*** - American sycamore
13. ***Carya texana*** - black hickory
14. ***Rhus aromatica*** - fragrant sumac

Acknowledgements

This version of the key has been developed and revised numerous times after extensive field use by dendrology students over the past 15 years. I wish to thank all who served as test subjects! I wish also to thank my co-author, Mr. Lance C. Jones. Lance served as my teaching assistant for several years and is now considered one of the leading taxonomists of east Texas. He also collected many of the specimens for the photography project. Other teaching assistants (Shaun Crook, Jared Laing, Carey Russell, Jon King, Andy McCrady and Amy Braig) have all provided input into the usefulness of the key. This reference key would not have been completed without the dedication and artistic ability of Bruce Lyndon Cunningham who is an artist, a forester, and a friend.

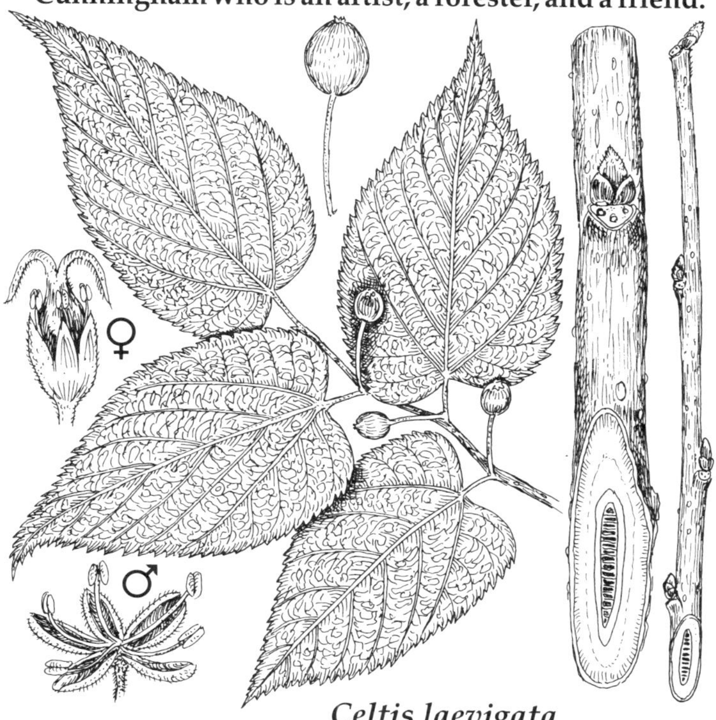

Celtis laevigata

MAP OF EAST TEXAS COUNTIES

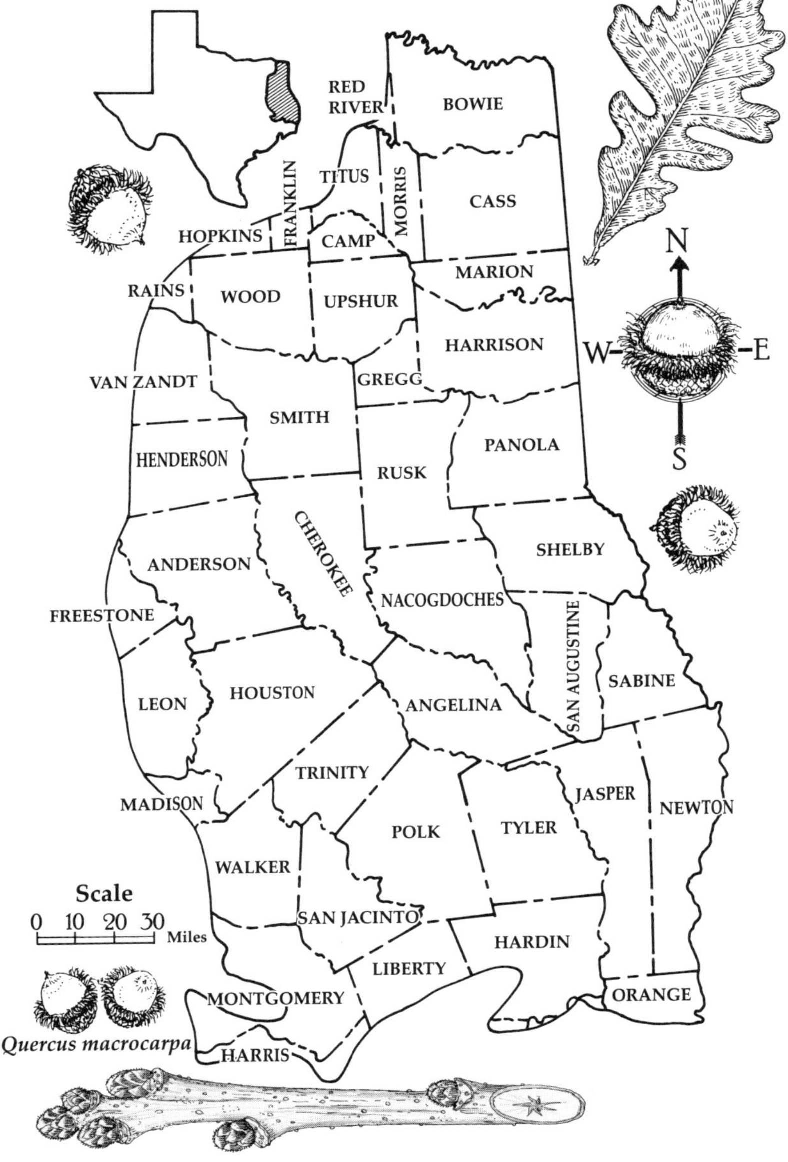

Quercus macrocarpa

TABLE OF CONTENTS

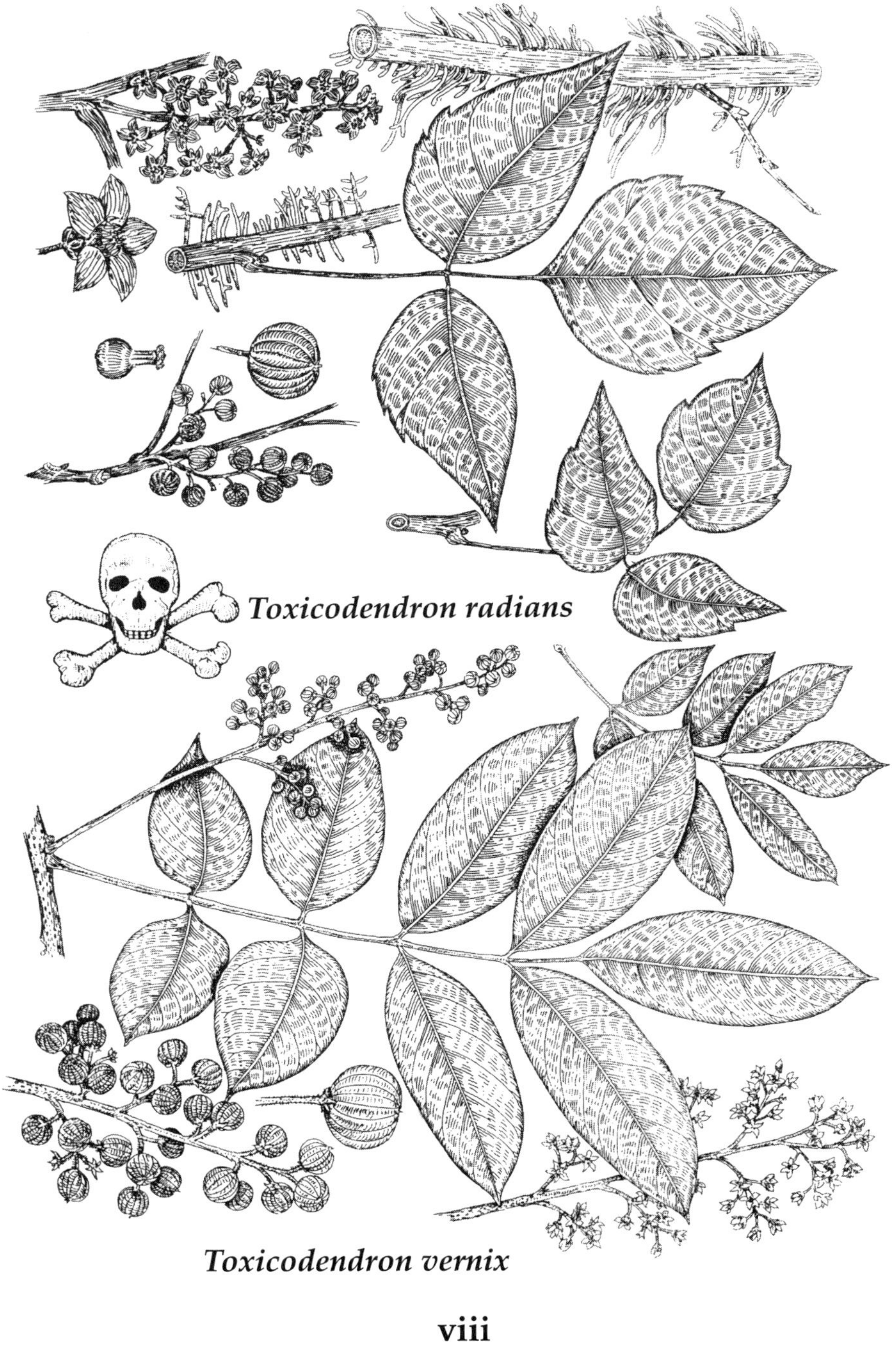
Toxicodendron radians
Toxicodendron vernix

Winter Key to Deciduous Woody Plants of East Texas

****Caution:** This key may cause you to handle poison-ivy or poison-sumac! Carefully handle vines that climb with rootlets along the vine, as this could be poison-ivy. Single stemmed plants in the understory of baygalls or seeps could be poison-sumac. Use caution handling such plants on these sites until the identification is correct. When you see this symbol, 

,go to the back cover to see the species in color.

1. Leaves are persistent
   (use an appropriate growing season key - see Page 86)
1. Leaves are deciduous. . . . .(2)

   2. Plant is a Woody vine . . . . **Group I**, Page 2
   2. Plant is a Shrub or Tree . . . .(3)

3. Lateral buds opposite or sub-opposite
   or whorled . . . .(4)
3. Lateral buds alternate on the twig. . . . . .(5)

   4. Vascular bundle scar one,
      appearing as a dot or straight,
      curved,or U-shaped line .. . .**Group II**, Page 4
   4. Vascular bundle scars three, in 3
      groups, or numerous, often closely
      spaced and arranged in a U-shape
      or as an ellipse. . . . .**Group III**, Page 6

5. Twigs armed with thorns,
   spines or prickles . . . . **Group IV**, Page 10
5. Twigs unarmed . . . .(6)

   6. Pith chambered . . . .**Group V**, Page 14
   6. Pith solid, either homogenous
      or diaphragmed. . . . .(7)

7. Twigs with true terminal bud. . . . . **Group VI**, Page 16
7. Twigs with false terminal bud
   (branch or twig scar evident). . . . .**Group VII**, Page 24

Group I. Woody Vines

1. Climbing by means of rootlets along the stem. . . . .(2)
1. Not climbing by means of rootlets along the stem . . . . (3)

2. Leaf scars alternate; buds densely hairy and without visible bud scales, leaf scars with 5 or more bundle scars; . . . . .poison-ivy, *Toxicodendron radicans* ⇨
2. Leaf scars opposite, bundle scars U-shaped to circular; vine tan brown; . . . . .trumpet-creeper, *Campsis radicans* ⇨

3. Vines armed, pith lacking, stem is solid except for scattered pores; . . . . greenbrier, *Smilax* spp. ⇨
3. Vines unarmed. . . . .(4)

4. Twining or scrambling stems . . . . (5)
4. Climbing by means of tendrils. . . . . (6)

5. Buds oblong and appressed; leaf scars raised; stem smooth and greenish-colored; . . . . Alabama supplejack, *Berchemia scandens* ⇨
5. Buds small and hairy with 3 bundle scars in elliptical-shaped leaf scars; pith large, white; . . . . .Carolina snailseed-vine, *Cocculus carolinus* ⇨

6. Tendrils and lateral buds on same side of twig; . . . . . .American buckwheat-vine, *Brunnichia cirrhosa* ⇨
6. Tendrils arise from same node as lateral buds but on opposite side. . . . .(7)

7. Pith white and continuous, never with dark woody partitions at the nodes; tendrils terminate in adhesive disks; . . . . .Virginia creeper, *Parthenocissus quinquefolia* ⇨
7. Pith brownish and continuous except for dark woody partitions at the nodes; tendrils do not terminate in adhesive disks; . . . .(8)

8. Tendrils are forked near the end; . . . . .summer grape, *Vitis aestivalis* ⇨
8. Tendrils are not forked near the end; . . . . .muscadine grape, *Vitis rotundifolia* ⇨

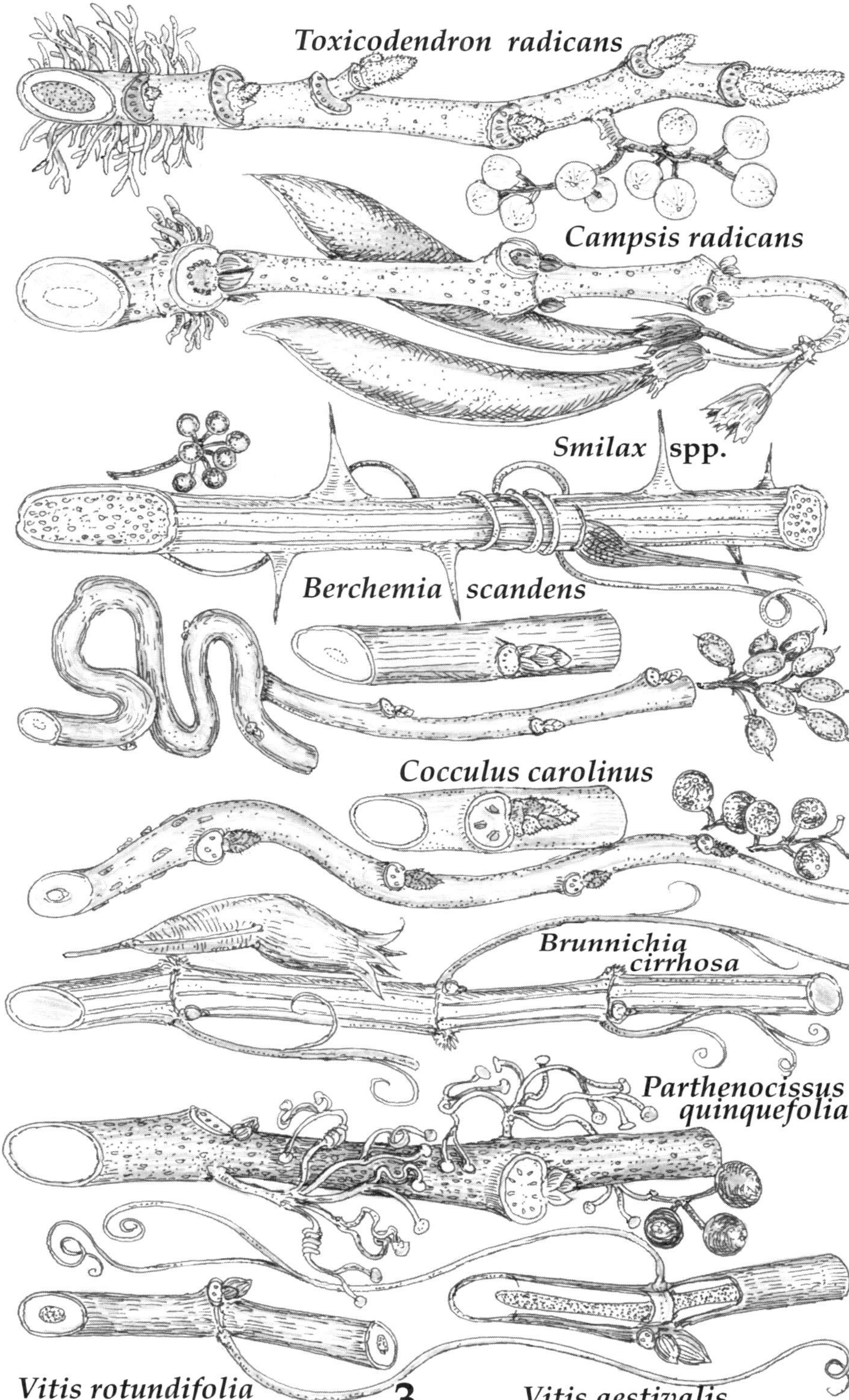

Toxicodendron radicans
Campsis radicans
Smilax spp.
Berchemia scandens
Cocculus carolinus
Brunnichia cirrhosa
Parthenocissus quinquefolia
Vitis rotundifolia
Vitis aestivalis

Group II. Twigs and Buds Opposite, Sub-opposite, or Whorled; Bundle Scar Single

1. Leaf scars 2-3 at each node (whorled); buds buried in bark of twigs; stipules persist through winter as 2 blackened "ears" on top edge of leaf scar; fruit rounded, persisting to late winter; . . . . common buttonbush, *Cephalanthus occidentalis* ⇨
1. Leaf scars usually only 2 per node (opposite, may be sub-opposite) . . . .(2)

    2. Twigs green, distinctly 4-sided, lenticels not warty; . . . . strawberry-bush, *Euonymus americanus* ⇨
    2. Twigs not green, twigs rounded (terete) in cross-section may be somewhat flattened. . .(3)

3. Leaf scars small and indistinct, torn, partially connected by ridges; bark on main stems shreddy; . . . . .coral-berry, *Symphoricarpos orbiculatus* ⇨
3. Leaf scars not as above . . . .(4)

    4. Leaf scars nearly circular; fruit stalks persistent throughout the winter, branched, resembling "elk antlers"; twigs scurfy-hairy; . . . . . . American beautyberry, *Callicarpa americana* ⇨
    4. Leaf scars U-shaped to half-round or shield-shaped. . . . .(5)

5. Lateral buds distinctly scaly . . . .(6)
5. Lateral buds indistinctly scaly, often are partially depressed into the twig. . . . .(9)

    6. Bundle scars numerous, arranged in a distinct circle; twigs large; (escaped exotic) . . . . .royal paulownia, *Paulownia tomentosa* ⇨
    6. Bundle scars not as above, small and indistinct or arranged in a U-shape . . . . . .(7)

7. Bundle scars arranged in a U-shape; lenticels very warty; leaf scars craterlike, often sub-opposite; twigs fairly stout;. . . . . . .fringetree, *Chionanthus virginicus* ⇨
7. Bundle scars small and indistinct; twigs very slender near tips. . . . .(8)

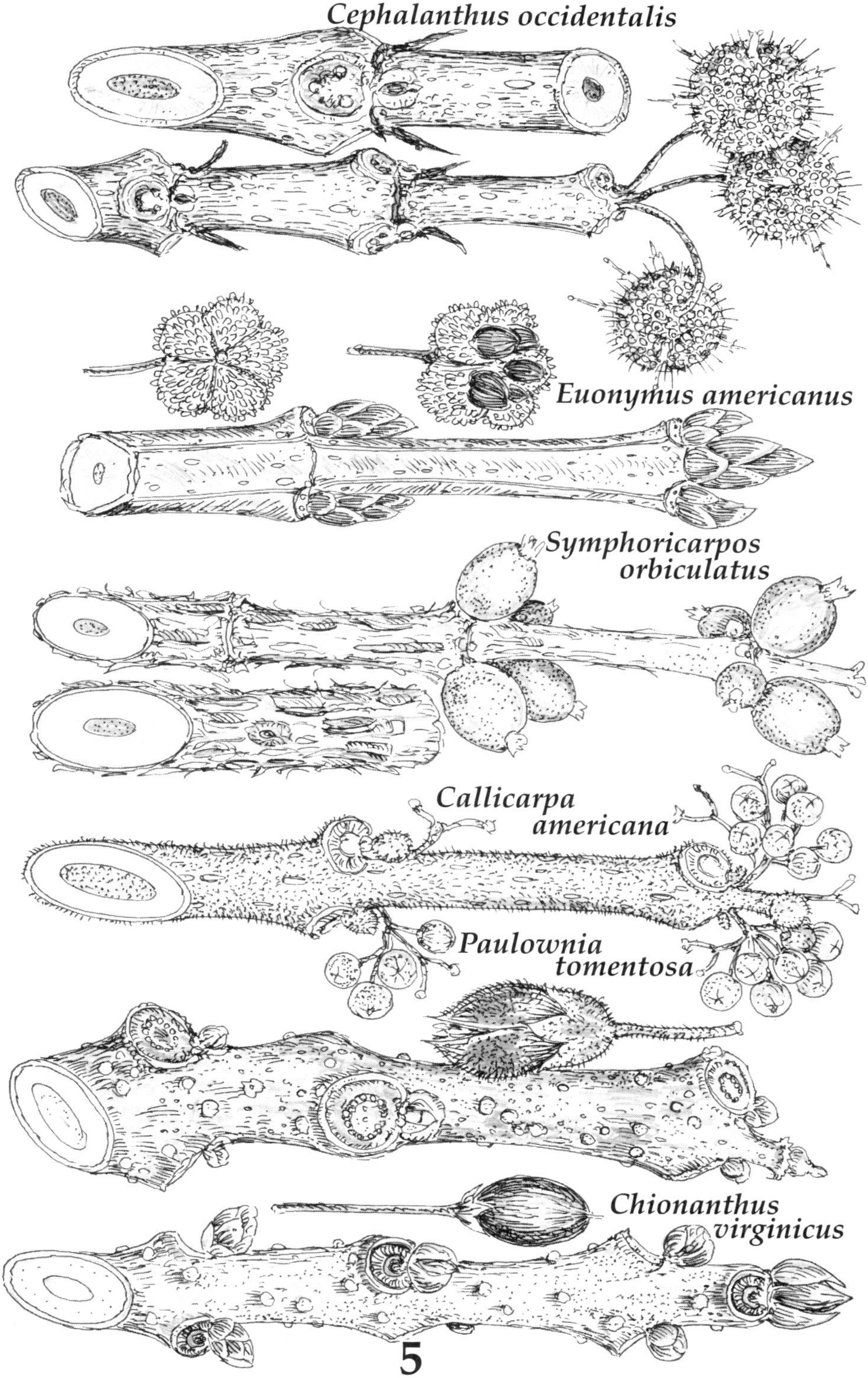
Cephalanthus occidentalis
Euonymus americanus
Symphoricarpos orbiculatus
Callicarpa americana
Paulownia tomentosa
Chionanthus virginicus

8. Twigs finely pubescent to slightly scabrous; plants of upland sites; ....upland forestiera, *Forestiera ligustrina* ⇨

8. Twigs essentially glabrous; plants native to riparian or wet sites; .... swamp-privet, *Forestiera acuminata* ⇨

9. Lateral buds are deeply indented into the leaf scar; leaf scar is notched, either U or V-shaped with the bud down in the notch; twigs glabrous; ....white ash, *Fraxinus americana* ⇨

9. Lateral buds located on top of leaf scar; leaf scar may be slightly notched at the top...............(10)

10. Stipule scars indistinct or lacking; mesic and bottomland sites; .... green ash, *Fraxinus pennsylvanica* ⇨

10. Stipule scars evident, appear as a ridge pointing down the twig; swampy sites or bog sites; .... Carolina ash, *Fraxinus caroliniana* ⇨

Group III. Twigs and Buds Opposite, Sub-opposite, or Whorled; Bundle Scars 3 or More

1. True terminal bud absent (i.e., a pseudo-terminal bud is present or there is no terminal at all).........(2)

1. True terminal bud present .... (4)

2. Leaf scars usually 3-whorled, craterlike, each leaf scar at slightly different locations at the node; bundles scars arranged in an ellipse; fruit is a long capsule (20-50 cm), persisting over winter; .... northern catalpa, *Catalpa speciosa* ⇨

2. Leaf scars not whorled, not craterlike.......(3)

3. Bundle scars 3-7, usually arranged in a curve or slight V-shape; pith large and spongy; plants commonly leaf out during warm periods during the winter; .... American elder, *Sambucus canadensis* ⇨

3. Bundle scars in 3 distinct groups; pith is 6-sided; top view of buds shows 4 distinct sides; .... red buckeye, *Aesculus pavia* ⇨

4. Bundle scars more than 3, may be in 3 groups .(5)

4. Bundle scars 3 ....(8)

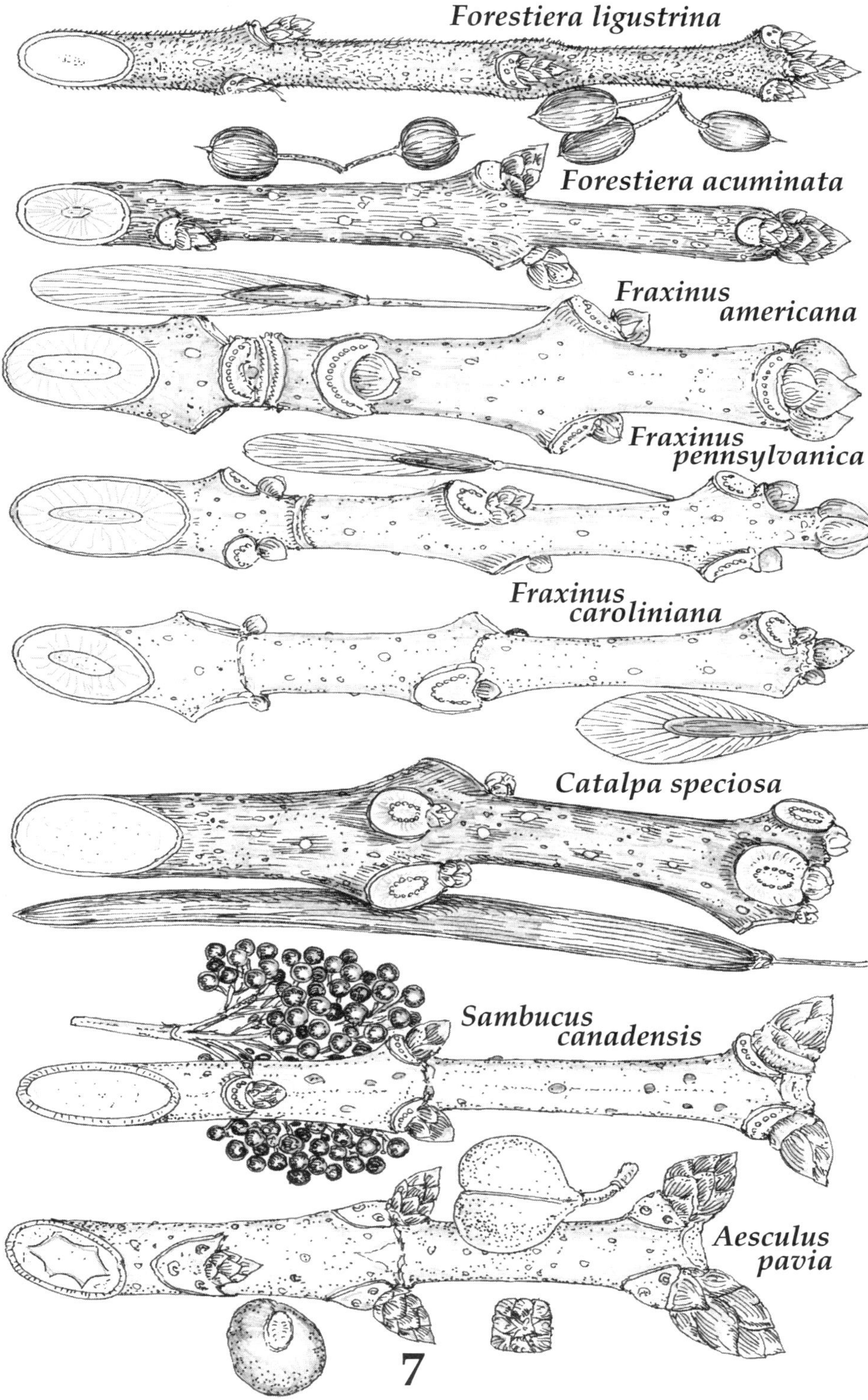
Forestiera ligustrina
Forestiera acuminata
Fraxinus americana
Fraxinus pennsylvanica
Fraxinus caroliniana
Catalpa speciosa
Sambucus canadensis
Aesculus pavia

5. Terminal buds large; top view of terminal bud shows it is 4-sided; pith is 6-sided; . . . . red buckeye, *Aesculus pavia* ⇨

5. Terminal buds not as above; pith is nearly circular. . . . . (6)

6. Lateral buds deeply indented into leaf scars, giving leaf scars a distinct U or V-shape; . . . . white ash, *Fraxinus americana* ⇨

6. Lateral buds not deeply indented into leaf scar, lateral buds are located on top of leaf scar. . . . .(7)

7. Stipule scars indistinct or lacking; mesic to bottomland sites;. . . . . green ash, *Fraxinus pennsylvanica* ⇨

7. Stipule scars evident, appear as a ridge connecting the leaf scars; swampy or bog sites; .Carolina ash, *Fraxinus caroliniana* ⇨

8. Outer scales of terminal bud 2, valvate, may be hairy and difficult to distinguish. .(9)

8. Outer scales of terminal bud more than 2, imbricate . . . .(13)

9. Terminal bud large, somewhat globose at base, covered with rusty brown scale-like hairs; twigs of season gray to nearly silver-colored; . . . . rusty blackhaw, *Viburnum rufidulum* ⇨

9. Terminal bud not as above . . . .(10)

10. Terminal bud flattened and elongated, covered with orange-brown hairs; species is only found in or adjacent to bogs or seeps; . . . . .possumhaw viburnum, *Viburnum nudum* ⇨

10. Terminal buds not greatly elongated or flattened; twigs of season highly colored (reddish-purple, reddish or green); leaf scars V- or U-shaped. . . .(11)

11. Twigs covered with fine hairs, red to reddish green, often scabrous; . . .rough-leaf dogwood, *Cornus drummondii* ⇨

11. Twigs essentially glabrous. (12)

12. Twigs purplish or dark bluish-green, glaucous; fruit a reddish colored drupe; twigs are telescoping; . . . . flowering dogwood, *Cornus florida* ⇨

12. Twigs green to reddish brown; buds brownish-hairy; habitat is saturated soils in bottomlands, ... stiff (English) dogwood, *Cornus stricta*. . . . . . . ⇨

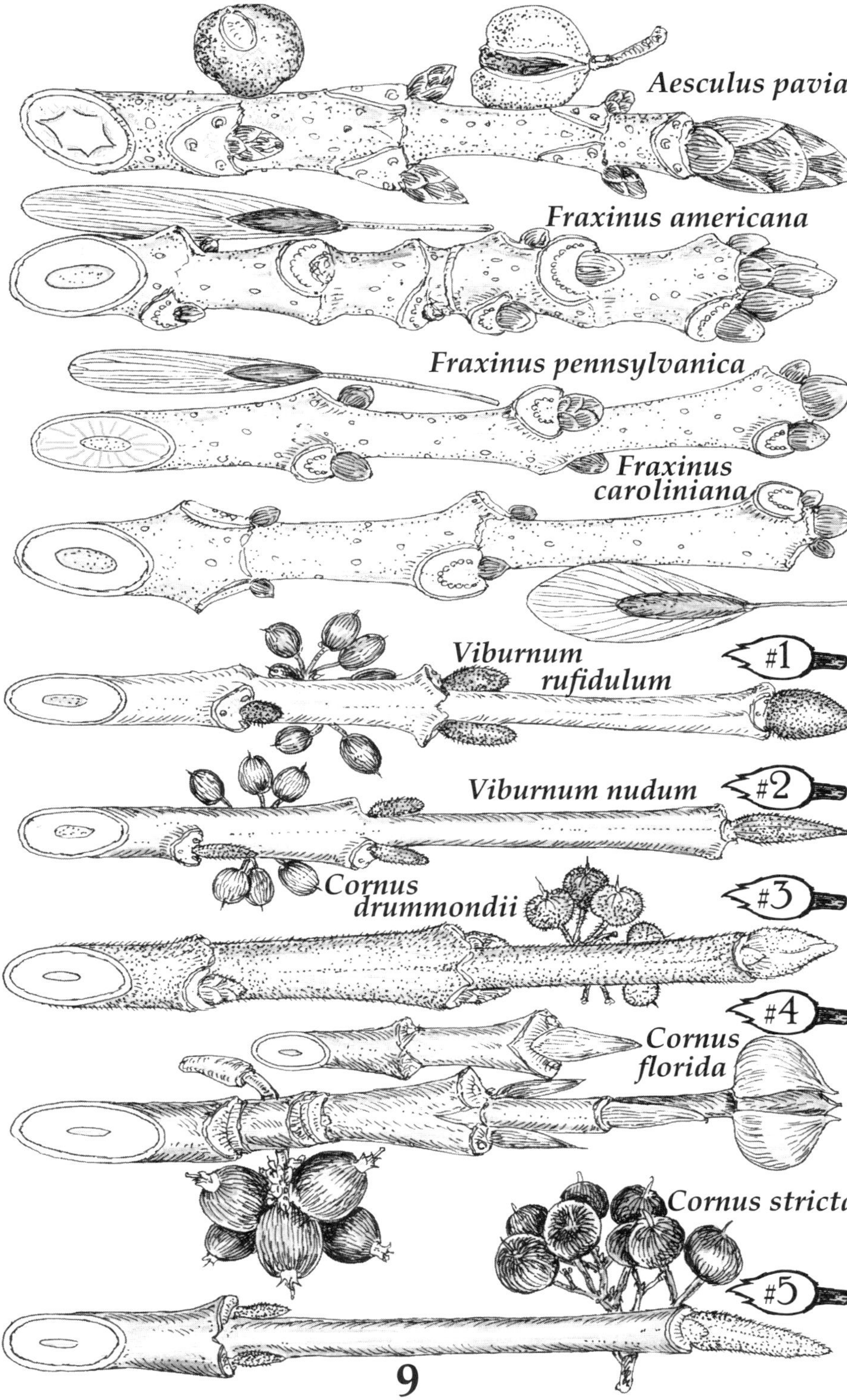
Aesculus pavia
Fraxinus americana
Fraxinus pennsylvanica
Fraxinus caroliniana
Viburnum rufidulum
#1
Viburnum nudum
#2
Cornus drummondii
#3
#4
Cornus florida
Cornus stricta
#5

13. Twigs bright green and shiny, very brittle; buds reddish purple and covered with a whitish pubescence, terminal often poorly developed; . . . . .boxelder, *Acer negundo* ⇨
13. Twigs not as above. . . . .(14) #6

14. Twigs scurfy hairy to tomentose; terminal bud not hairy except near base of bud; . . . . arrow-wood viburnum, *Viburnum dentatum* ⇨
14. Twigs essentially glabrous, shiny brown or red. . . . .(15)

15. Buds brown or blackish-brown, sharp-pointed, bud scales 6 or more; twigs glossy, buff or reddish-brown; . . . . Florida maple, *Acer barbatum* ⇨
15. Buds and twigs reddish or orange-brown, about 4 visible scales. . . . .(16) #7

16. Twigs bright chestnut brown; inner bark with a rank and disagreeable odor; lateral buds straight-sided at base; bud scales ciliate on upper edges; . . . . silver maple, *Acer saccharinum* ⇨ #8
16. Twigs red and lustrous; buds scales red; inner bark without rank odor;. . . . . .red maple, *Acer rubrum* ⇨ #9

Group IV. Twigs Armed with Thorns, Spines, or Prickles

1. Twigs and stems green; buds appear at the notch where the twigs branch; . . . . trifoliate-orange, *Poncirus trifoliata* ⇨
1. Twigs and stems not green . . . . (2)

2. Spines distributed over the surface of the twigs and stems . . . .(3)
2. Spines or thorns solitary or paired at the nodes. . . . .(4)

3. Plant usually consists of one erect stem, unbranched; bundle scars 5 or more; leaf scars exceptionally large and extending more than halfway around the stem; . . . . devil's walkingstick, *Aralia spinosa* ⇨
3. Plants branched, not as above; bundle scars 3, leaf scars round or triangular; buds almost black; bark on main bole bearing conical projections that end in a spine; prickles on twigs almost black; . . . . .Hercules-club, *Zanthoxylum clava-herculis* ⇨

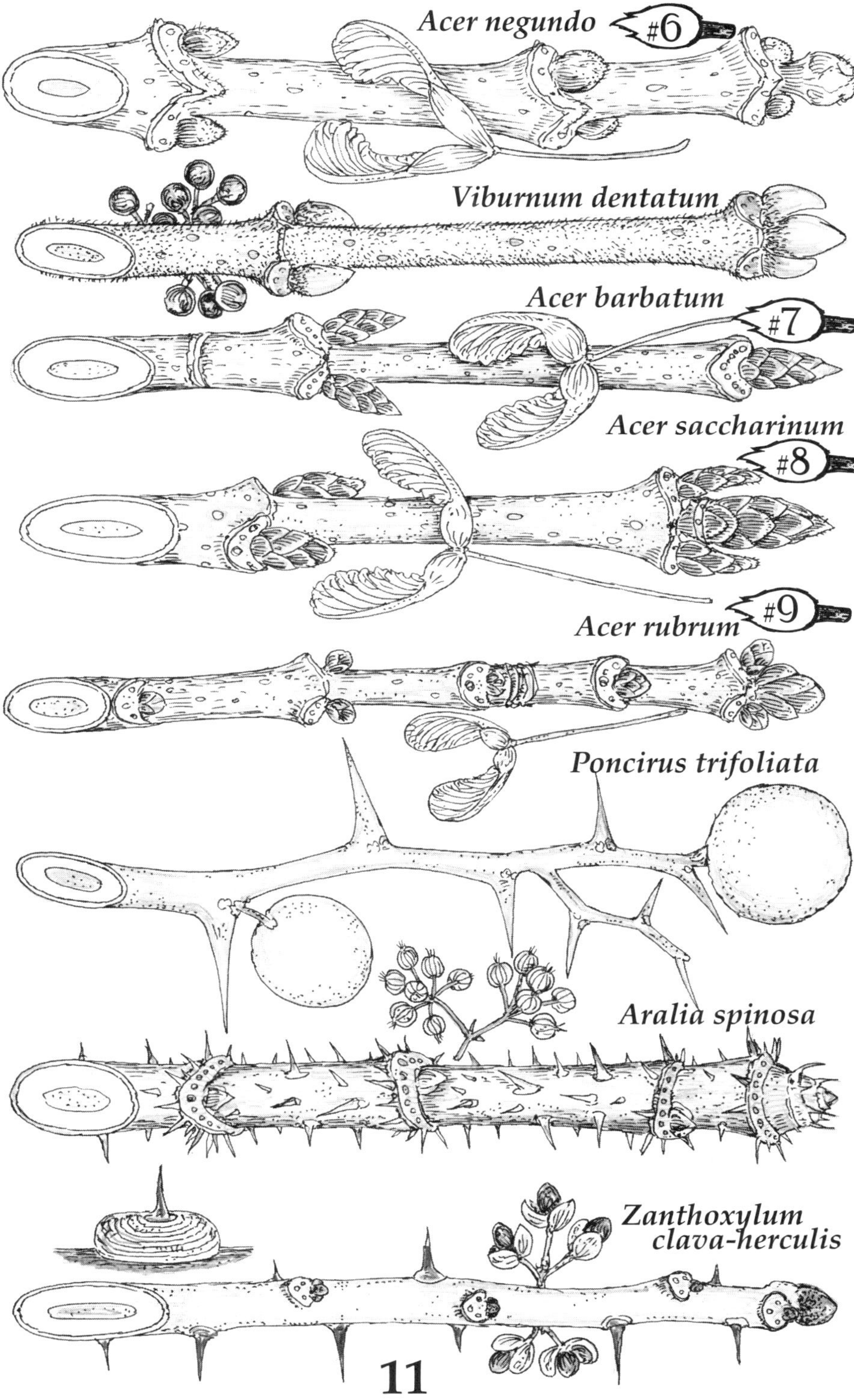
Acer negundo
#6
Viburnum dentatum
Acer barbatum
#7
Acer saccharinum
#8
#9
Acer rubrum
Poncirus trifoliata
Aralia spinosa
Zanthoxylum clava-herculis

4. Spines paired at the nodes; terminal bud absent; lateral buds submerged above the leaf scar;
.... black locust, ***Robinia pseudoacacia*** ⇨

4. Spines or thorns solitary at the nodes. ....(5)

5. Thorns usually 2-3 branched, red to purplish-colored; fruit is a legume .... (6)

5. Thorns are not branched, fruit is not a legume ....(7)

6. Seed pods linear-oblong with many seeds imbedded in tissue; mostly upland sites;
...............honeylocust, ***Gleditsia triacanthos*** ⇨

6. Seed pods oval with 1-3 seeds not imbedded in tissue, mostly bottomland sites that are frequently flooded
..................waterlocust, ***Gleditsia aquatica*** ⇨

7. True terminal bud present; bud scales dark red; thorns usually long and slender;
.... hawthorns, ***Crataegus*** **spp.** ⇨

7. Twigs with a false terminal bud; small, spur-like branches often present. ....(8)

8. Bud scales brown, buds depressed, partially imbedded in the twig; bark of mature tree orange-brown; sap milky;
.... Osage-orange, ***Maclura pomifera*** ⇨

8. Bud scales rusty tomentose; bark dark brown and anastomosing; leaf scars often covered with hairs; spinescent twigs usually present;
.............chittamwood, ***Sideroxlyon lanuginosa*** ⇨

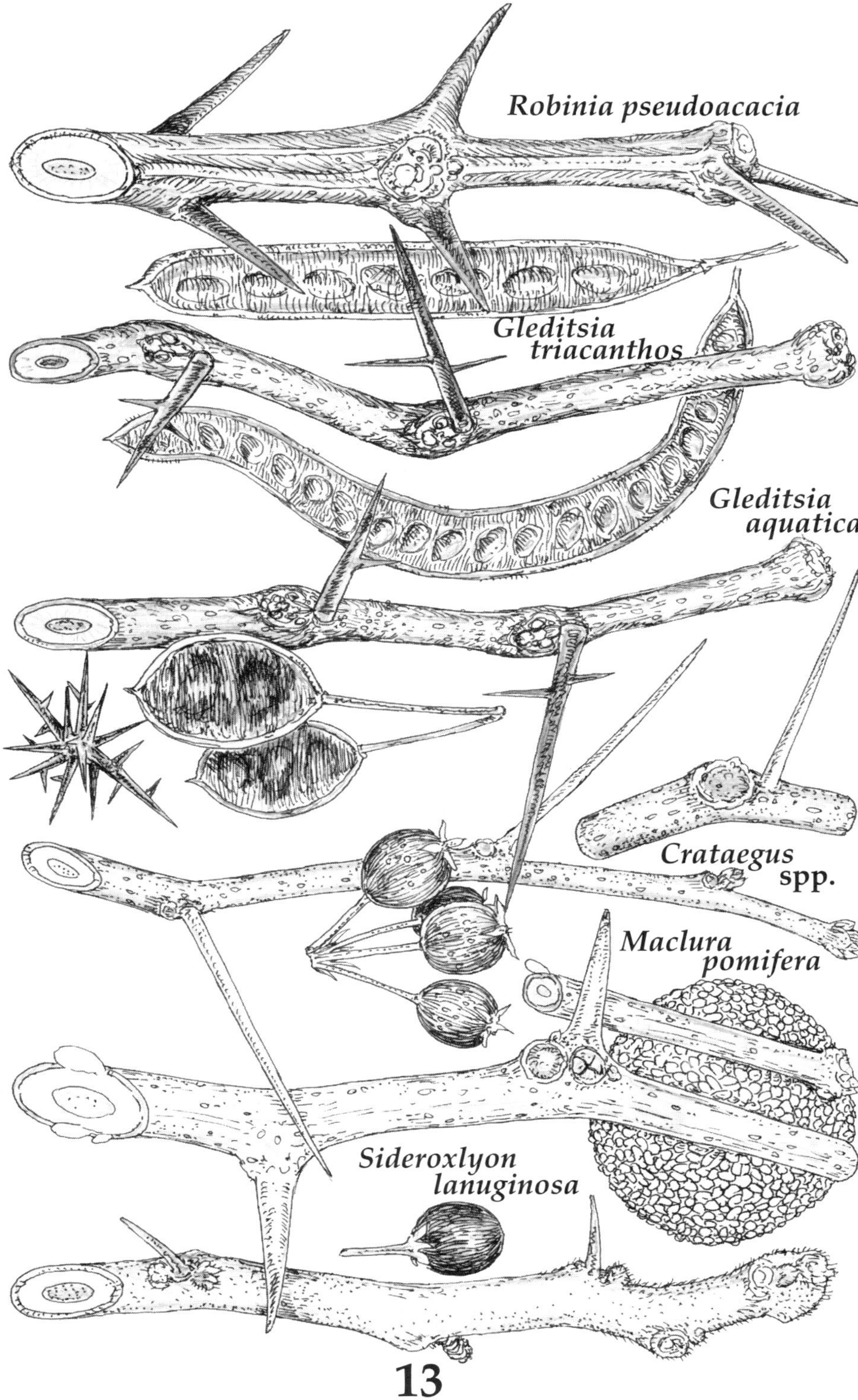
Robinia pseudoacacia
Gleditsia triacanthos
Gleditsia aquatica
Crataegus spp.
Maclura pomifera
Sideroxlyon lanuginosa

Group V. Twigs Unarmed, Pith Chambered

1. **Solitary bundle scar, occasionally resembling numerous scars closely spaced . . . .(2)**
1. **Bundle scars 3 or more than 3 (may be in 3 groups) . . . .(4)**

    2. **Buds apparently covered with only 2 scales that greatly overlap; branch scar usually prominent; bark divided into distinctive scaly blocks on mature trees; . . . . common persimmon,** *Diospyros virginiana* ⇨
    2. **Buds obviously with more than 2 scales . . . .(3)**

3. **Buds usually clustered near the end of the twig, true terminal bud present; leaves often persistent; . . . . sweetleaf,** *Symplocos tinctoria* ⇨
3. **Buds not clustered near the end, pseudo-terminal bud solitary near the tip of twig; leaves totally deciduous; . . . . Carolina silverbell,** *Halesia carolina* ⇨

    4. **Twigs green or greennish-red sometimes; . . . . .Virginia sweetspire,** *Itea virginica* ⇨
    4. **Twigs not green . . . .(5)** #10

5. **Leaf scars broad and conspicuous, shield-shaped or 3-lobed; bundle scars large and in 3 groups, usually appearing as 3 small horseshoes; buds large and grayish pubescent; superposed buds often present; . . . . .black walnut,** *Juglans nigra* ⇨
5. **Leaf scars small; pith small and often chambered only near the nodes . . . .(6)**

    6. **Buds obviously imbricate scaled; twigs zigzag and 2-ranked; . . . . .sugarberry,** *Celtis laevigata* ⇨
    6. **Buds densely hairy with orange-brown hairs; terminal elongated and flattened; small shrub on dry, upland sites; . . . . .dwarf pawpaw,** *Asimina parviflora* ⇨

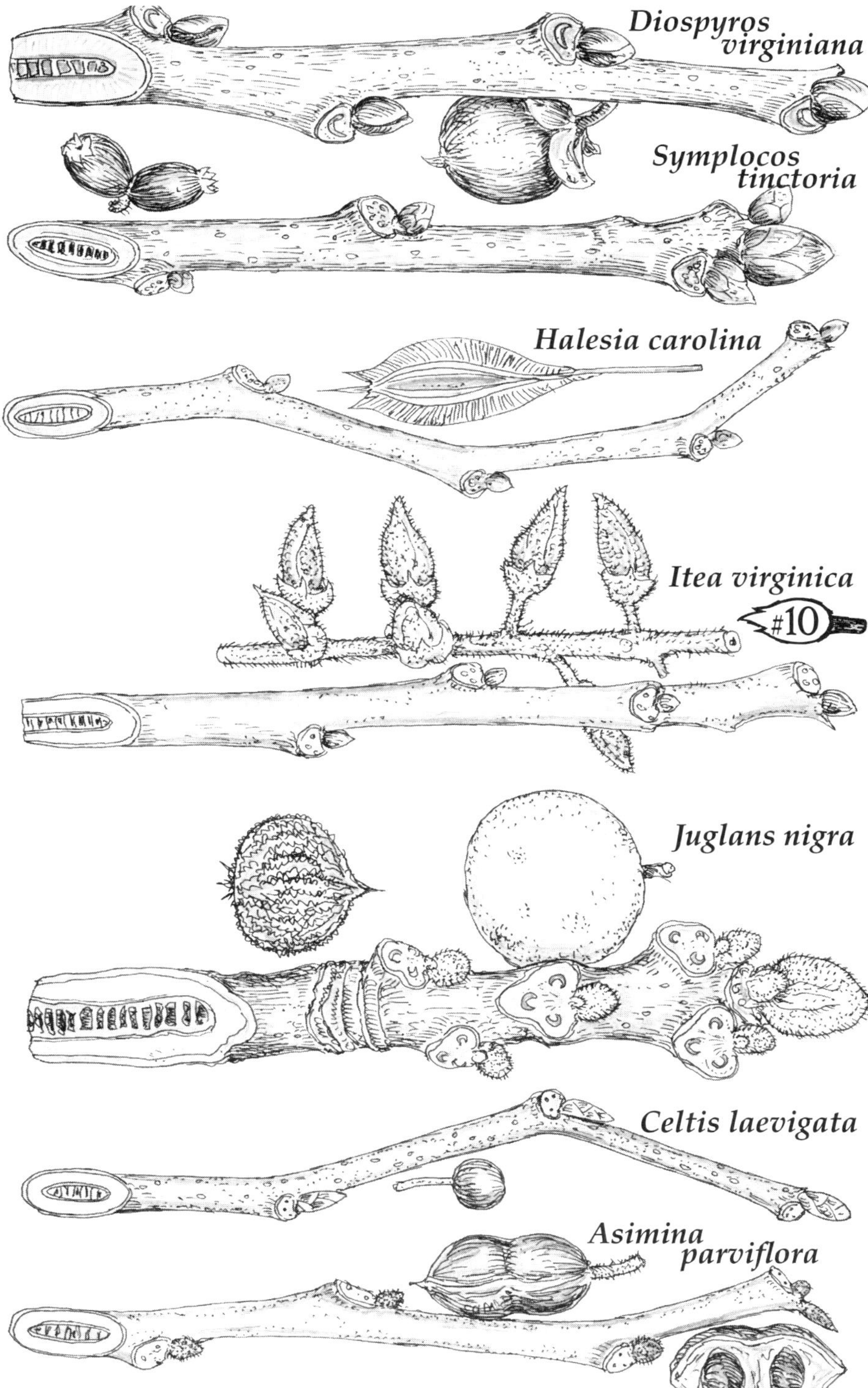
Diospyros virginiana
Symplocos tinctoria
Halesia carolina
Itea virginica
#10
Juglans nigra
Celtis laevigata
Asimina parviflora

Group VI. Twigs Unarmed, Pith Solid, True terminal Bud Present

1. Pith diaphragmed . . . .(2)
1. Pith homogenous or rarely spongy. . . . .(6)

2. Stipule scars appearing as a line encircling the twig, bundle scars numerous. . . . . (3)
2. Stipule scars lacking or if present, not encircling the twig; bundle scars 3 to 5. . . . .(4)

3. Twigs green to black-mottled with green; buds silky whitish pubescent; leaves tardily deciduous; #11 . . . . .sweetbay, *Magnolia virginiana*. . . . . . ⇨
3. Twigs brown to purplish-brown; bud scales glabrous, yellowish to straw colored, distinctly valvate;. . . . . . . yellow-poplar, *Liriodendron tulipifera*. . . . . . ⇨ #12 (not native)

4. Buds distinctly scaly; bundle scars 3; bud scales dark, purplish-black; bark of mature trees distinctly fissured and broken into square blocks; #13 . . . . black tupelo, *Nyssa sylvatica*. . . . . . . ⇨
4. Buds appear naked due to dense covering of velvety hairs; terminal bud longer than laterals and somewhat flattened. . . . . (5)

5. Terminal bud covered with rusty-colored hairs; shrubs to small tree size growing in low woods and bottomlands; . . . . . . .pawpaw, *Asimina triloba* ⇨
5. Terminal bud covered with orange to golden brown hairs, small; shrub of dry sandy upland sites;. . . . dwarf pawpaw, *Asimina parviflora*. . . . . .⇨ #14

6. Twigs brownish, lateral buds submerged beneath the bark; 1-3 globose buds with greenish valvate scales on terminal end of twigs; twigs also have round pits with a small knob extending from center of pits, edges of pits irregular; mature trees with distinct buttress swell and knees arising from roots close to surface; . . . . baldcypress, *Taxodium distichum* ⇨
6. Twigs and buds not as above. . . . . . . (7)

7. Stipule scar forming a complete ring around the twig; terminal buds silky whitish pubescent; leaves #11 tardily deciduous;. . . . . sweetbay, *Magnolia virginiana* ⇨
7. Stipule scars lacking or if present, not completely encircling the twig. . . . .(8)

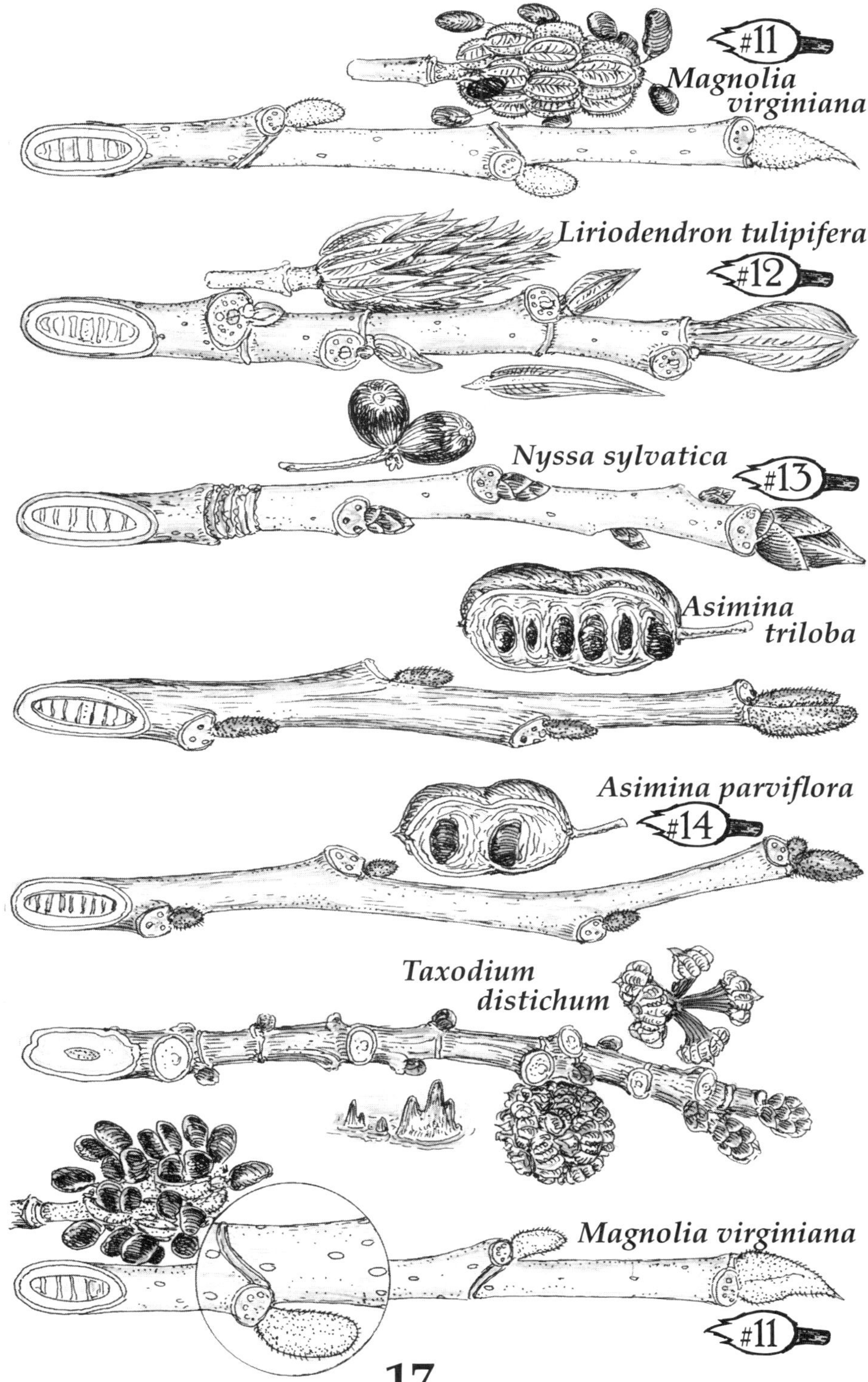
#11
Magnolia virginiana
Liriodendron tulipifera
#12
Nyssa sylvatica
#13
Asimina triloba
Asimina parviflora
#14
Taxodium distichum
Magnolia virginiana
#11

8. Leaf scar contains a single bundle scars; new twigs are bright green, end of branches turn upward; terminal bud with 3-4 visible scales, each scale having a single central midrib along its back; . . . . sassafras, *Sassafras albidum*. . . . . . . #15 ⇨

8. Twigs and buds scales not as above. . . . . (9)

9. Buds naked (dense hairs also causes the naked appearance) . . . . (10)

9. Buds scaly . . . . (13)

10. Buds wholly or partially covered with dense red or orange hairs, (See #5. on page 16) . . . . pawpaw or dwarf pawpaw, *Asimina* spp. ⇨

10. Buds greenish yellow, tan-colored, or brown to black. . . . . (11)

11. Stipule scars lacking, leaf scars large and shield-shaped; bundle scars in 3 groups; pith usually 5-angled; . . . . hickories, *Carya* spp. ⇨ (See *Carya* spp. key on page 36)

11. Stipule scars or persistent stipules present, leaf scars small and not lobed . . . . (12)

12. Buds greenish yellow to tan, sometimes with a pair of small, almost inconspicuous scales near the base; superposed buds (accessory) present; stipules deciduous, stipule scars evident; . . . . witch-hazel, *Hamamelis virginiana*. . . . . . #16 ⇨

12. Buds brownish, entirely naked; stipules usually persistent; . . . . Carolina buckthorn, *Frangula caroliniana*. . . . . #17 ⇨

13. Bundle scar one, appearing as a dot, straight line or curved line. . . . . (14)

13. Bundle scars 2 or more . . . . (19)

14. Stipules are small, may be persistent or deciduous . . . . (15)

14. Stipules or stipule scars lacking, terminal bud present or deciduous. . . . . (16)

Sassafras albidum #15

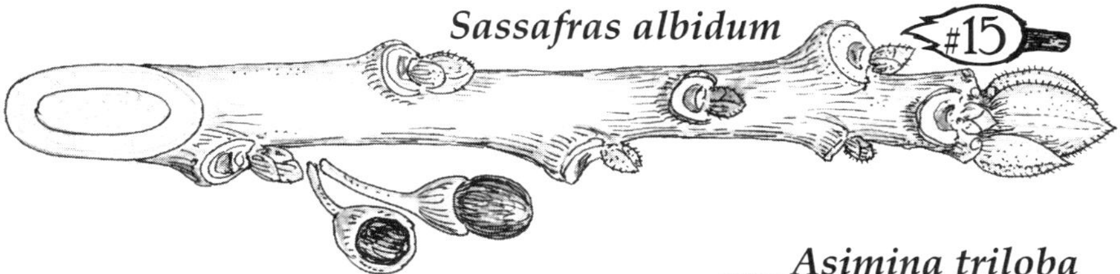

Asimina triloba

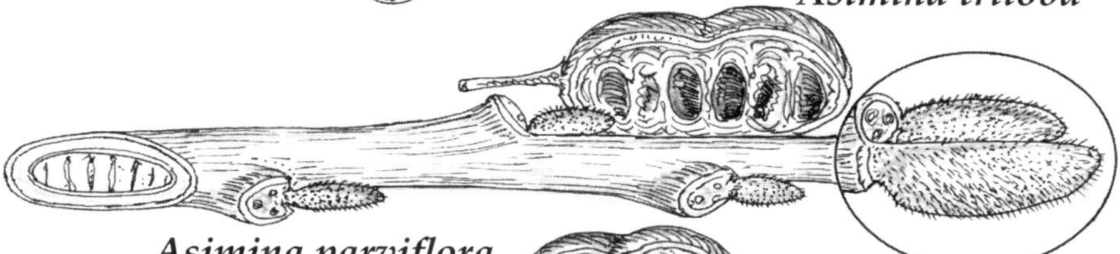

Asimina parviflora

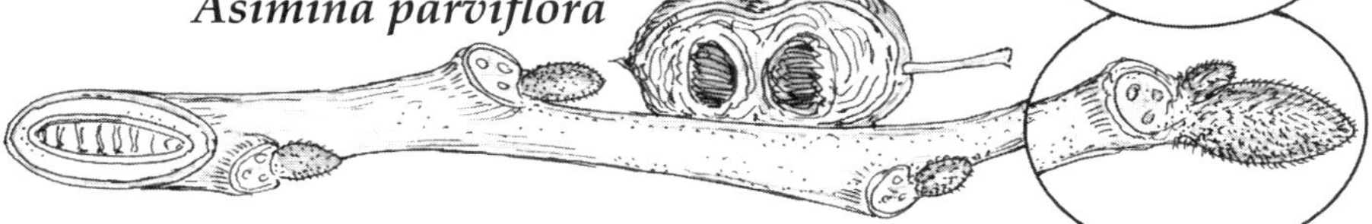

Carya illinoinensis

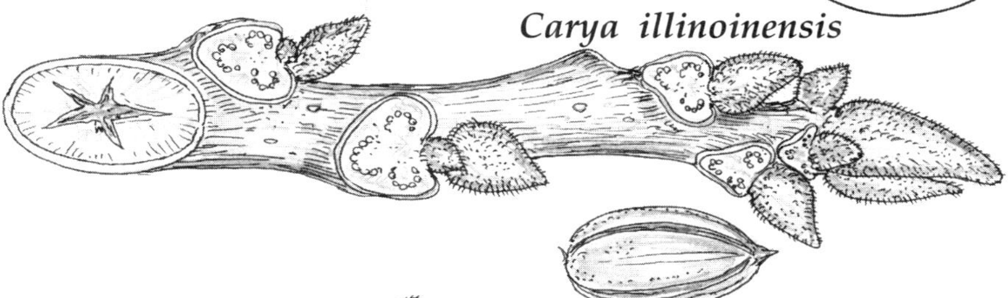

Carya cordiformis

Hamamelis virginiana #16

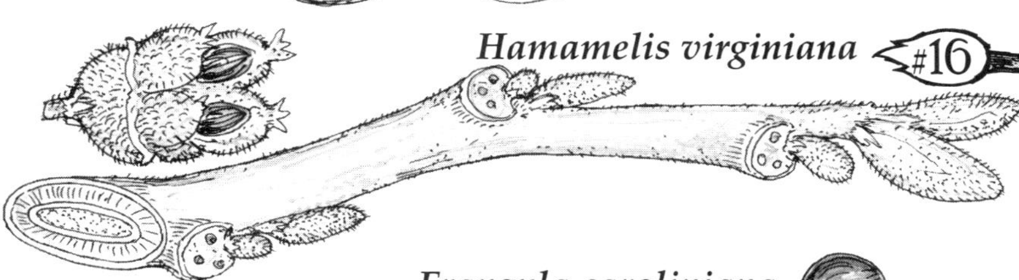

Frangula caroliniana #17

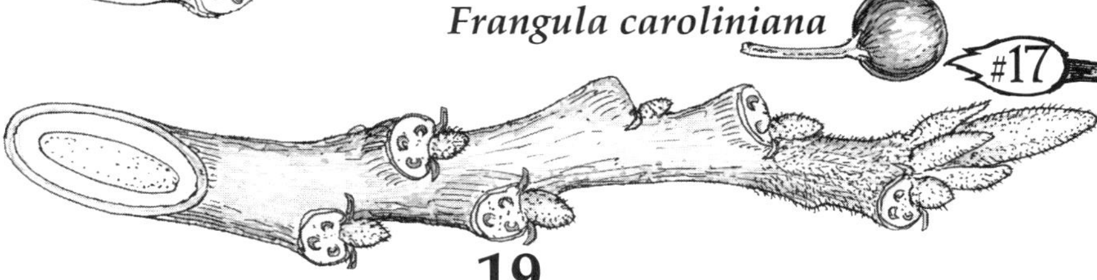

15. Stipules often persist as minute spines, particularly near terminal end of twig; buds are small, terminal poorly developed; bundle scar always single; . . . . .deciduous holly (possumhaw), *Ilex decidua* ⇨

15. Stipules and buds not as above; terminal bud obvious; bundle scar mostly single but sometimes indistinctly 3 in number; . . . . Carolina buckthorn, *Frangula caroliniana* ⇨

16. Twigs and buds with fine hairs. . . . . (17)

16. Twigs not hairy, twigs often greenish. (18)

17. Hairs appear to be stellate or are set in clusters or tufts;. . . . . .sweet pepper-bush, *Clethra alnifolia* ⇨

17. Hairs do not appear stellate or clustered; branches often appear whorled near top of plant; buds and leaf scars often clustered or whorled at top of twig; . . . . .Texas azalea, *Rhododendron oblongifolium* ⇨

18. Shrub less than 2 m tall, never arborescent; . . . . .blueberry, *Vaccinium* spp. ⇨

18. Shrub greater than 2 m tall, often attaining a diameter exceeding 25 mm; . . . . tree sparkleberry, *Vaccinium arboreum* ⇨

19. Stipule scars long and narrow, nearly encircling the twig; terminal bud usually 2-3 cm long, resembles "cigar"; . . . . .American beech, *Fagus grandifolia* ⇨

19. Twigs and buds not as above. . . . .(20)

20. Cross section shows pith to be 3 to 5 angled, triangular, 4-sided, or star-shaped . . . .(21)

20. Cross section shows pith to be circular or nearly so. . . . . (27)

21. Buds distinctly stalked, pith is triangular; . . . . hazel alder, *Alnus serrulata* ⇨

21. Buds are sessile, not stalked; pith is 4-sided to star-shaped. . . . .(22)

22. Basal scale of lateral bud greenish and clasping the base of the bud, buds resinous and sticky when crushed; twigs usually ridged, not terete; . . .eastern cottonwood, *Populus deltoides* ⇨

22. Bud scales not as above . . . . (23)

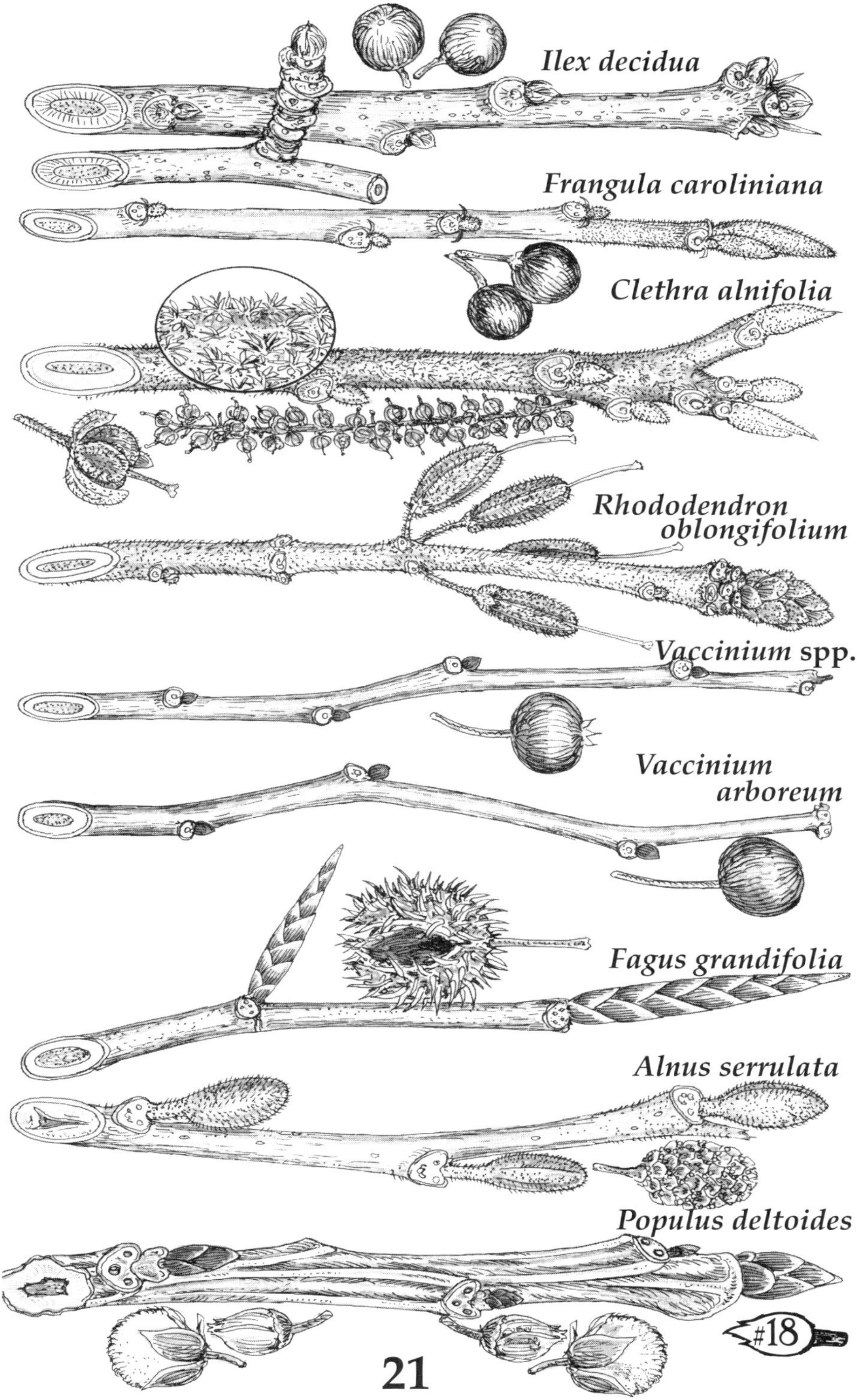

Ilex decidua
Frangula caroliniana
Clethra alnifolia
Rhododendron oblongifolium
Vaccinium spp.
Vaccinium arboreum
Fagus grandifolia
Alnus serrulata
Populus deltoides
#18

23. Buds usually clustered near the end of the twig; fruit is an acorn; . . . . .oaks, *Quercus* spp. ⇨
(see separate key to species on page 38)

23. Buds not clustered near ends of twigs; fruit not an acorn. . . . . (24)

24. Leaf scars lobed, large and shield-shaped; bundle scars numerous and forming small horseshoes; . . . . true hickories, *Carya* spp. ⇨
(see separate key to species, page 36)

24. Leaf scars half-round to elliptical, bundle scars 3 or in 3 groups . . . .(25)

25. Twigs with a bitter, strongly disagreeable taste; main stem and older twigs with distinctive horizontal lenticels; . . . . .black cherry, *Prunus serotina* ⇨

25. Twigs not having bitter taste . . . .(26)

26. Bundle scars nearly ringlike with dark centers; twigs greenish to red-brown, often with corklike excrescences; buds with more than 3 imbricate scales; . . . . . sweetgum, *Liquidambar styraciflua*. . . . ⇨ #19

26. Bundle scars not as above, twigs lacking corky wings; buds 2-3 scaled; terminal bud sometimes deciduous; nutlike fruit enclosed in a spiny bur, rarely present due to immaturity of stems due to disease that kills boles; often found as a sprout in a clump of dead stems; . . . . chinkapin, *Castanea pumila* ⇨

27. Buds conical or oblong. . . . . (28)

27. Buds globose . . . .(29)

28. Stipules or stipule scars present; twigs and stems with conspicuous horizontal lenticels; twigs may have disagreeable, bitter taste or odor; . . . . plums, *Prunus* spp. ⇨

28. Stipule scars absent; twigs wooly; terminal buds distinctly larger than laterals; bud scales red; fruits red; natural habitat is swampy areas or seeps; . . . . .red chokeberry, *Pyrus arbutifolia* ⇨

29. Lateral buds with 2 distinct scales, middle bundle scar distinctly V-shaped; . . . . .water tupelo, *Nyssa aquatica* ⇨

29. Bud scales reddish, more than 2 visible; twigs and stems with stems with conspicuous horizontal lenticels, may taste slightly bitter; . . . . plums, *Prunus* spp. ⇨

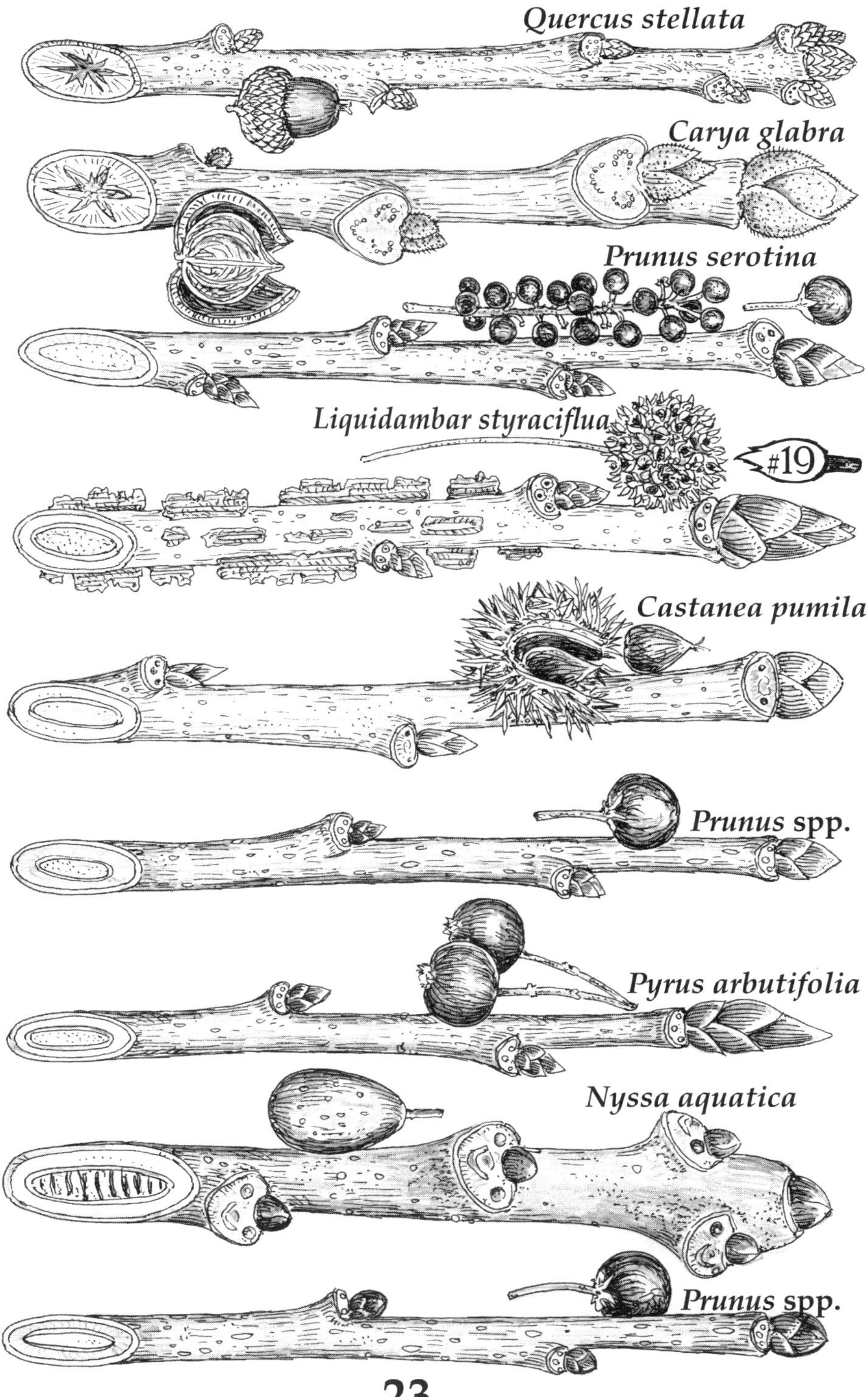
Quercus stellata
Carya glabra
Prunus serotina
Liquidambar styraciflua
#19
Castanea pumila
Prunus spp.
Pyrus arbutifolia
Nyssa aquatica
Prunus spp.

Group VII. Twigs Unarmed, Pith Solid, Pseudo-terminal Bud

1. Buds, especially laterals apparently lacking, obscured by or submerged above or behind the leaf scar;. . . . . (2)
1. Buds are eviden. . . . .(5)
(false terminal, laterals, or both are evident)

    2. Lateral buds partially submerged but often appearing as a round, small out-growth; (see Group IV - Page12) . . . . .honeylocust or waterlocust, *Gleditsia* spp. ⇨
    2. Lateral buds nearly completely submerged. . . . .(3)

3. Twigs small, lateral buds usually completely hidden behind raised leaf scar; small multi-stemmed shrub on dry upland sites;. . . . . .fragrant sumac, *Rhus aromatica* ⇨
3. Twigs stout. . . . .(4)

    4. Twigs green to blackish-green; #20 . . . . .sophora, *Sophora affinis* ⇨
    4. Twigs brown to gray-brown; usually armed with paired stipular spines at the nodes;. . . . . . black locust, *Robinia pseudoacacia* #21 ⇨

5. Leaf scars lacking (what appears to be leaf scars do not have bundle scars), buds very small, rounded, appearing partly submerged in the twig; mature trees distinctly buttress-swelled;. . . . . baldcypress, *Taxodium distichum* ⇨
5. Leaf scars present . . . .(6)

    6. Stipule scars form a closed ring around the twig, twigs very distinctly zigzag; leaf scars encircle the lateral buds; . . . . American sycamore, *Platanus occidentalis* ⇨
    6. Stipule or stipule scars lacking or if present, do not encircle the twig . . . . .(7)

7. Lenticels conspicuous and elongated horizontally (must be verified on older branches or main bole), lenticels on twigs may appear as dots. . . . . . .(8)
7. Lenticels appearing as dots, as warty structures, elongated vertically or are inconspicuous. . . . .(10)

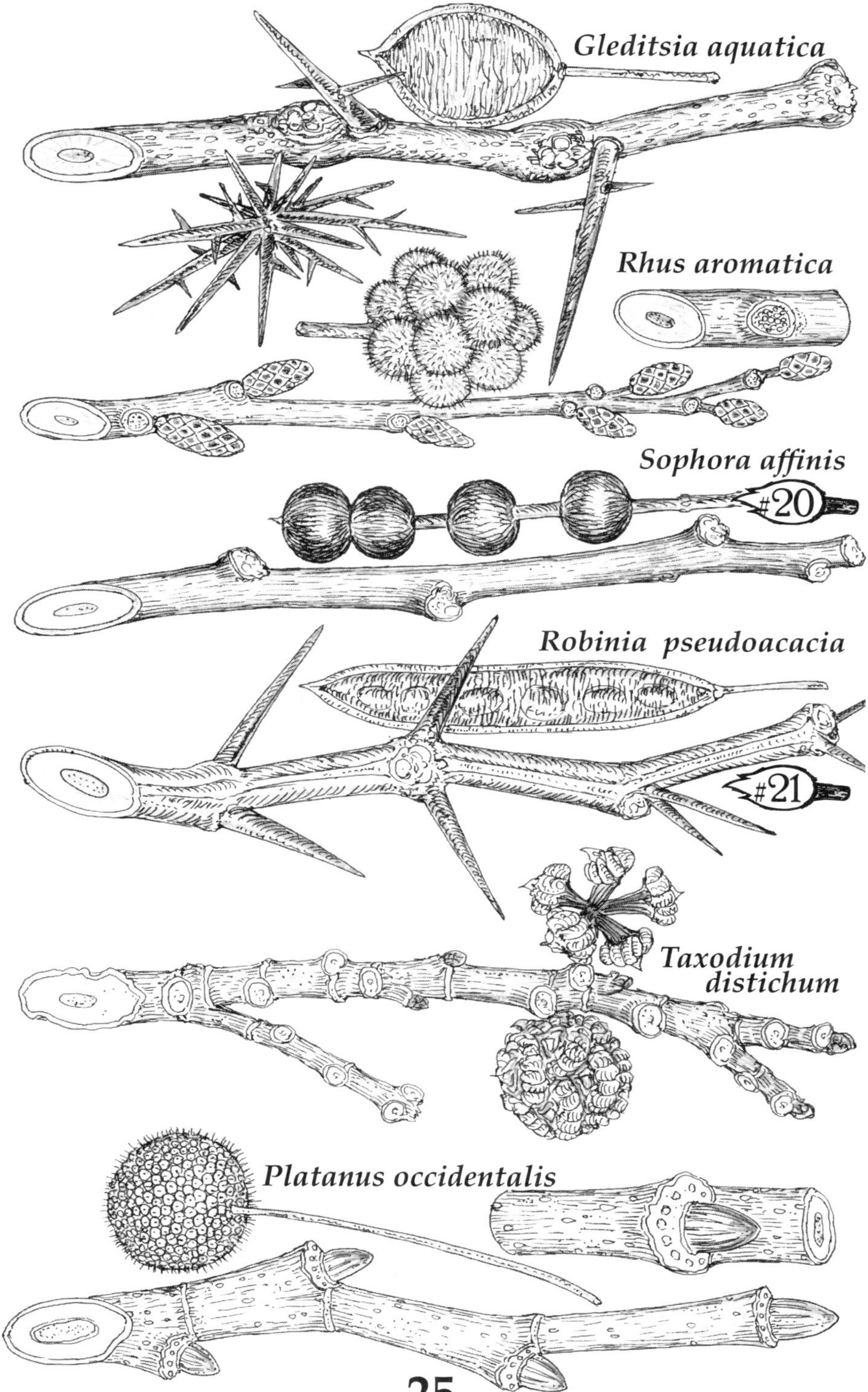
Gleditsia aquatica
Rhus aromatica
Sophora affinis
#20
Robinia pseudoacacia
#21
Taxodium distichum
Platanus occidentalis

8. Twigs large and stout, greenish to blackish-green, leaf scars distinctly 3 lobed; lateral buds covered with grayish pubescence; #22
. . . . .Chinaberry, *Melia azedarach*. . . . . . ⇨
8. Twigs, buds, and leaf scars not as above. . .(9)

9. Bark on main stem usually papery or very shreddy, peeling to reveal salmon-pink colored inner bark; bundle scars 3, middle scar is noticeably larger than the two bundle scars on the outer edges; usually growing as a riparian species; . . . . .river birch, *Betula nigra* ⇨
9. Bark never papery or shreddy; twigs and buds often have a bitter and disagreeable taste;
. . . . plums, *Prunus* spp. ⇨

10. Two spherical flower (accessory) buds usually present as collateral buds at each node; . . . . spicebush, *Lindera benzoin* ⇨
10 Collateral buds not present . . . . (11)

11. Buds appear naked, often due to dense pubescence covering the bud scales. . . . . (12)
11. Buds valvate or imbricate scaled or a single caplike scale covers the bud . . . .(18)

12. Bundle scars one, often torn and indistinct; superposed buds commonly present; present;. . . . . American snowbell, *Styrax americana* ⇨
12. Bundle scars 3, arranged in 3 groups, numerous . . . . (13)

13. Plant is commonly a single, unbranched stem; found only in seeps (baygall sites: sweetbay is dominant overstory; numerous fern species present in the understory);
. . . . poison-sumac, *Toxicodendron vernix* ⇨
13. Combination of plant habit and habitat not as above. . . . . (14)

14. Sap from freshly cut twig is milky; pith may be spongy. . . . .(15)
14. Sap not milky. . . . .(16)

15. Twigs glabrous and smooth, glaucous; . . . . smooth sumac, *Rhus glabra* ⇨
15. Twigs tomentose to pubescence; leaf scars U-shaped with 5-11 scattered bundle scars;. . . . . shining sumac, *Rhus copallina* ⇨

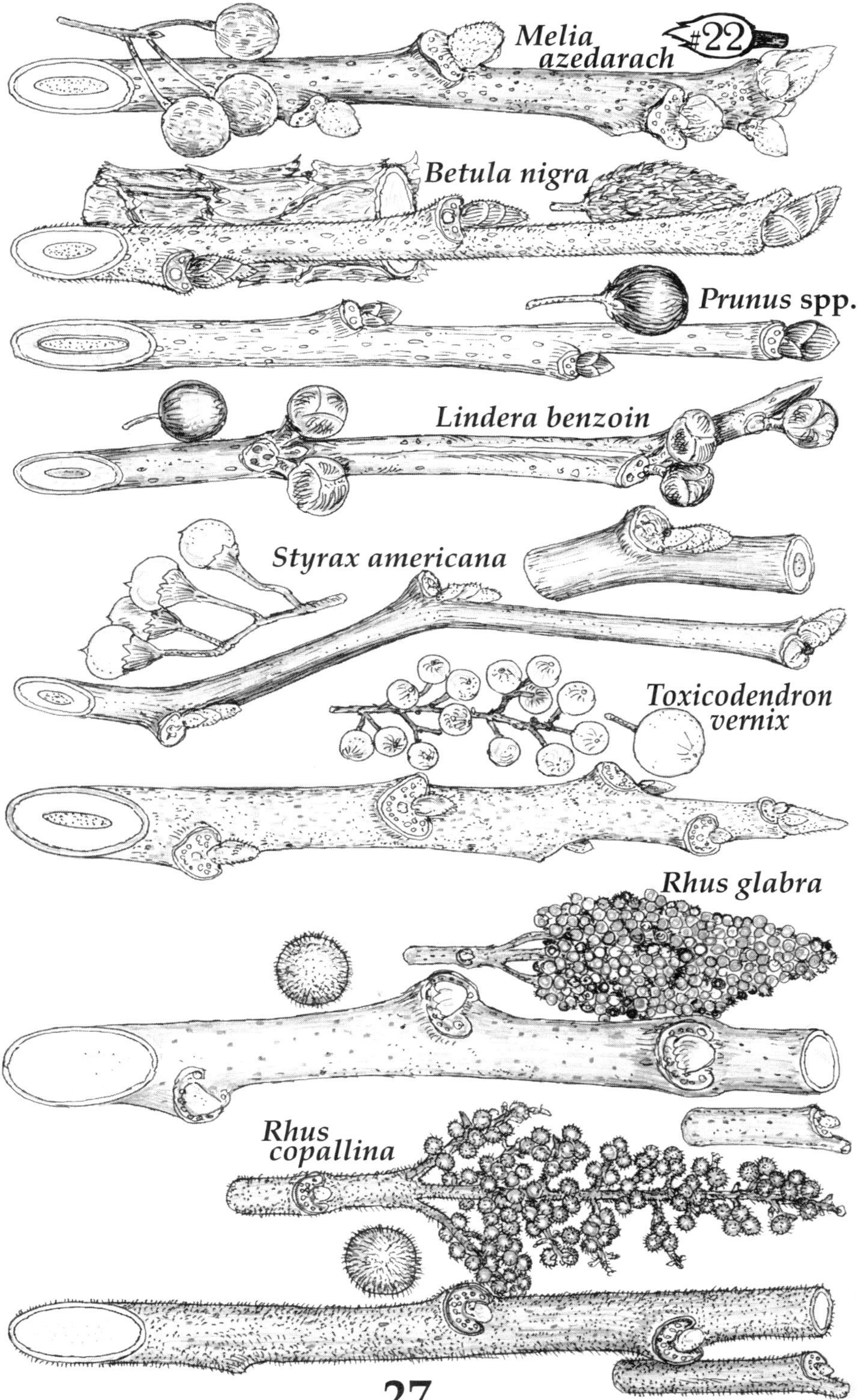
#22
Melia azedarach
Betula nigra
Prunus spp.
Lindera benzoin
Styrax americana
Toxicodendron vernix
Rhus glabra
Rhus copallina

16. Leaf scars horseshoe-shaped, bundle scars appear as 3 dots; leaf scars often nearly encircle the buds,. . . . . . . .common hoptree, *Ptelea trifoliata* ⇨
16. Leaf scars 3-lobed to shield-shaped; bundle scars U-shaped . . . .(17)

17. Twigs greenish to blackish-green; #22
. . . . Chinaberry, *Melia azedarach*. . . . . . . ⇨
17. Twigs light gray colored; superposed buds present, lowest one indented into the top of the leaf scar;
. . . . soapberry, *Sapindus drummondii* ⇨

18. Bud scale single, caplike, twigs separate easily at joints; leaf scars V-shaped with 3 bundle scars; stipule scars slit-like and inconspicuous; . . . . .black willow, *Salix nigra* ⇨
18. Bud scales and twigs not as above. . . . . (19)

19. Buds distinctly stalked; bud scales red to lavender-color; small conelike catkins (1 cm) persisting over winter;. . . .hazel alder, *Alnus serrulata*. . . . ⇨ #23
19. Buds sessile, not stalked. . . . .(20)

20. One bundle scar . . . .(21)
20. Bundle scars 2 or more . . . .(26)

21. Bundle scar in shape of a flattened U. . . . .(22)
21. Bundle scar small dot near center of leaf scar . . . .(24)

22. Twigs not zigzag; multi-stemmed small shrub, less than 2 m tall; twigs green; up to 10 cm or more of the top of the stems die back in winter but remain attached; bases of dry capsules usually persist into winter, resembling minature cups and saucers; roots dark red;
. . . . New Jersey tea, *Ceanothus americanus* ⇨
22. Twigs distinctly zigzag . . . .(23)

23. Buds covered by 2 greatly overlapping blackish-colored scales; pith occasionally chambered; bark of mature trees broken into squared blocks, giving the appearance of alligator skin;
. . . . .common persimmon, *Diospyros virginiana* ⇨
23. Buds with more than 2 scales, scales tan to brownish, pubescent; twigs splotchy brown; pith usually chambered;. . . . . Carolina silverbell, *Halesia carolina* ⇨

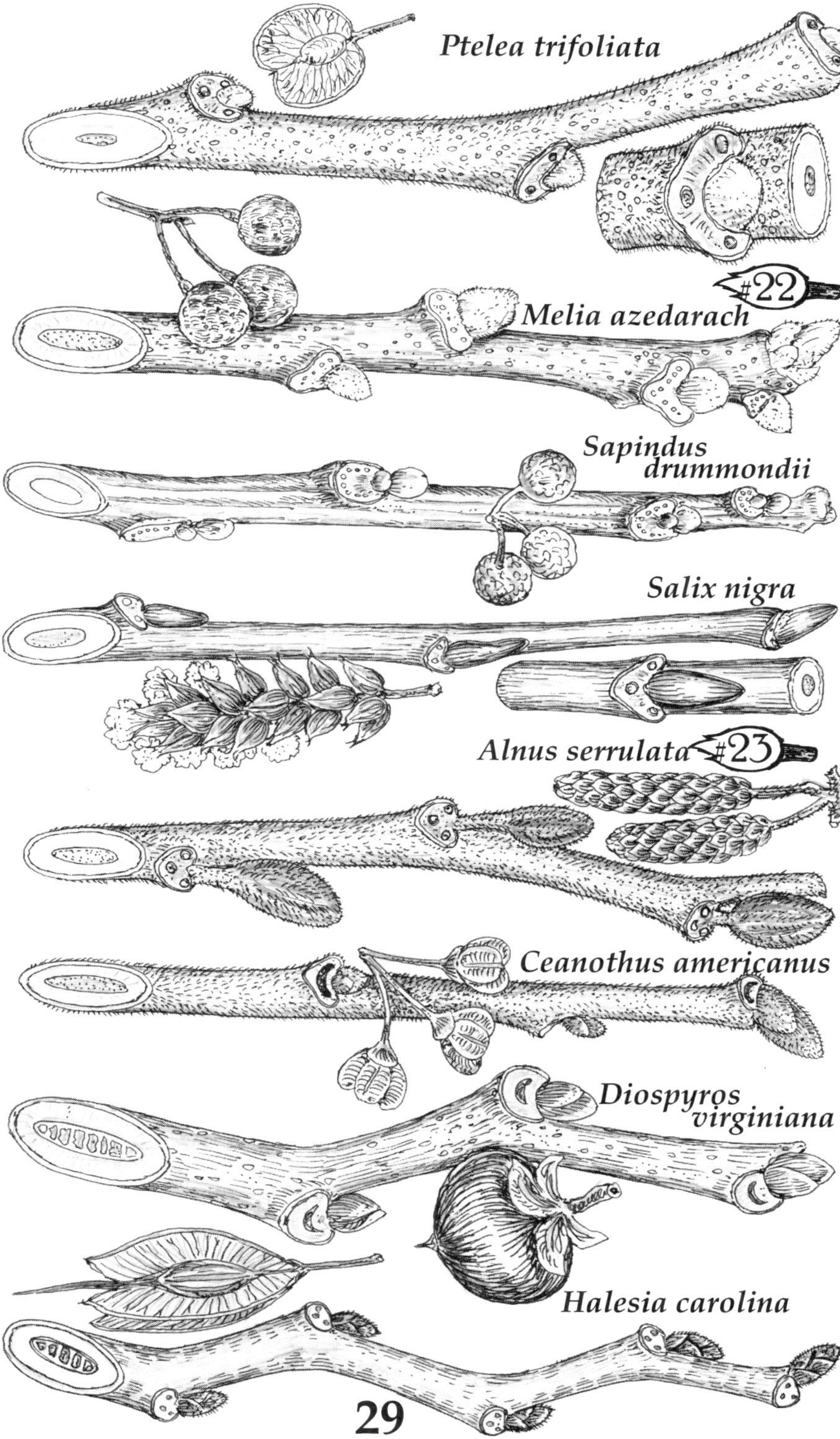
Ptelea trifoliata
#22
Melia azedarach
Sapindus drummondii
Salix nigra
Alnus serrulata
#23
Ceanothus americanus
Diospyros virginiana
Halesia carolina

24. Dry capsules usually persist into winter; shrub with distinctive striated bark, light gray in color; small, 5-valved capsules usually persist into winter; 2 distinct imbricate bud scales, reddish on one side of the bud and greenish on the other side; #24
. . . . . .he-huckleberry lyonia, *Lyonia ligustrina*. . . . . ⇨

24. Fruit generally absent in winter; fruit is a bluish-drupe . . . .(25)

25. Shrub less than 2 m tall, never arborescent;
. . . . blueberry, *Vaccinium* spp. ⇨

25. Shrub greater than 2 m tall, diameter often exceeding 25 cm;
. . . . .tree sparkleberry, *Vaccinium arboreum* ⇨

26. Twigs green, with fine groove running the length of the twig; bundle scars small; leaves partly persistent;
. . . . .eastern baccharis, *Baccharis halimifolia* ⇨

26. Twigs not green; leaves deciduous . . .(27)

27. Stipules or stipule scars lacking. . . . . (28)

27. Stipules or stipule scars present. . . . .(32)

28. Leaf scars with 3 bundle scars . . . .(29)

28. Leaf scars contain more than 3 bundle scars; twigs may be armed . . . .(30)

29. Leaf scars triangular shaped to 3-lobed; bundle scars distinctly raised; . . . . .silktree, *Albizia julibrissin* ⇨

29. Leaf scars not as above; twigs red-brown to purple with many small lenticels; flower buds may be present as collateral buds; #25
. . . . .eastern redbud, *Cercis canadensis*. . . . . . . ⇨

30. Lateral buds solitary or collateral, often borne on short, spurlike branches; bud scales wooly to tomentose;
. . . . .chittamwood, *Sideroxylon lanuginosa* ⇨

30. Lateral buds minute, sunken or submerged within leaf scar . . . .(31)

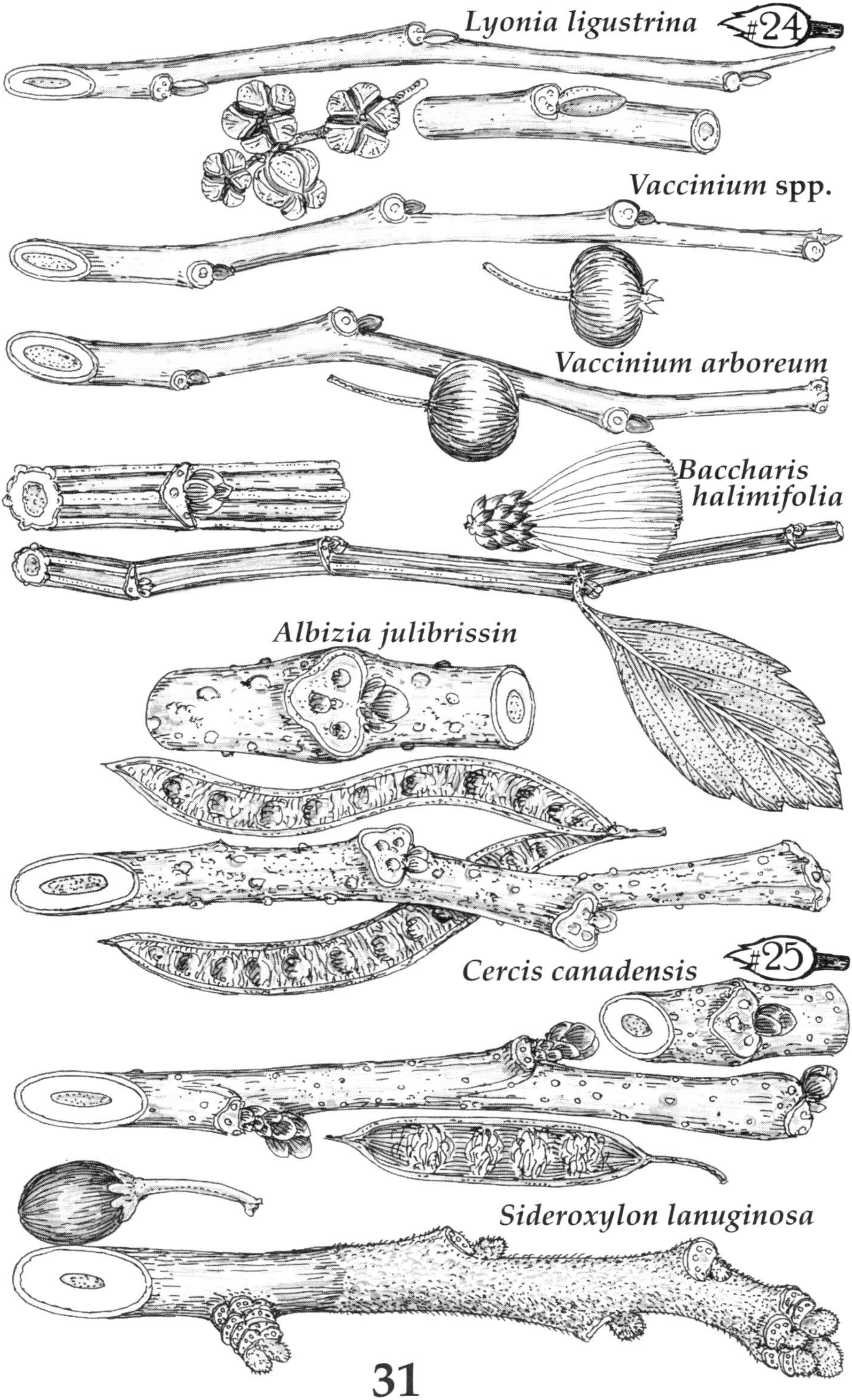
Lyonia ligustrina
#24
Vaccinium spp.
Vaccinium arboreum
Baccharis halimifolia
Albizia julibrissin
#25
Cercis canadensis
Sideroxylon lanuginosa

31. Twigs usually armed with branched thorns;
. . . . .honeylocust or waterlocust, *Gleditsia* spp. ⇒
31. Twigs large; young plants with few branches; pith very spongy, tan to brownish; bundle scars irregularly shaped and numerous, arranged in V-shape along edge of leaf scar; escaped exotic;
. . . . tree-of-heaven, *Ailanthus altissima* ⇒

32. Leaf scars and buds distinctly 2-ranked; numerous small twigs appear 2-ranked; buds closely appressed for nearly entire length of bud; pith occasionally intermittently chambered near the nodes; lenticels elongated vertically; bark on boles of mature trees with raised corky projections;
. . . . . sugarberry, *Celtis laevigata* ⇒
32. Leaf scars and buds not 2-ranked . . . . . .(33)

33. Bud 2-4 scaled. . . . . (34)
33. Buds with more than 4 scales . . . . (37)

34. Buds and twigs greenish-red to bright red, buds always 2-scaled, obliquely sessile, asymmetric;
. . . . . basswood, *Tilia* spp. ⇒
34. Twigs orange-brown to brown, buds symmetrical. . . . .(35)

35. Stipules deciduous, stipule scars evident; buds at first grayish tomentose but later turning reddish; often found as a sprout among clump of dead sprouts;
. . . . . .chinkapin, *Castanea pumila* ⇒
35. Stipules persistent as raised portions on sides of leaf scar . . . .(36)

36. Buds brown, somewhat triangular with indentation near the base on both sides; twigs with distinct small ridges running down from sides of leaf scars;
. . . . crapemyrtle, *Lagerstroemia indica* ⇒
(ornamental exotic)
36. Buds pale orange-brown; ends of twigs die back up to 30 to 45 cm; twigs do not have ridges running down from sides of leaf scars; escaped exotic;
. . . . .Chinese tallowtree, *Sapium sebiferum* ⇒

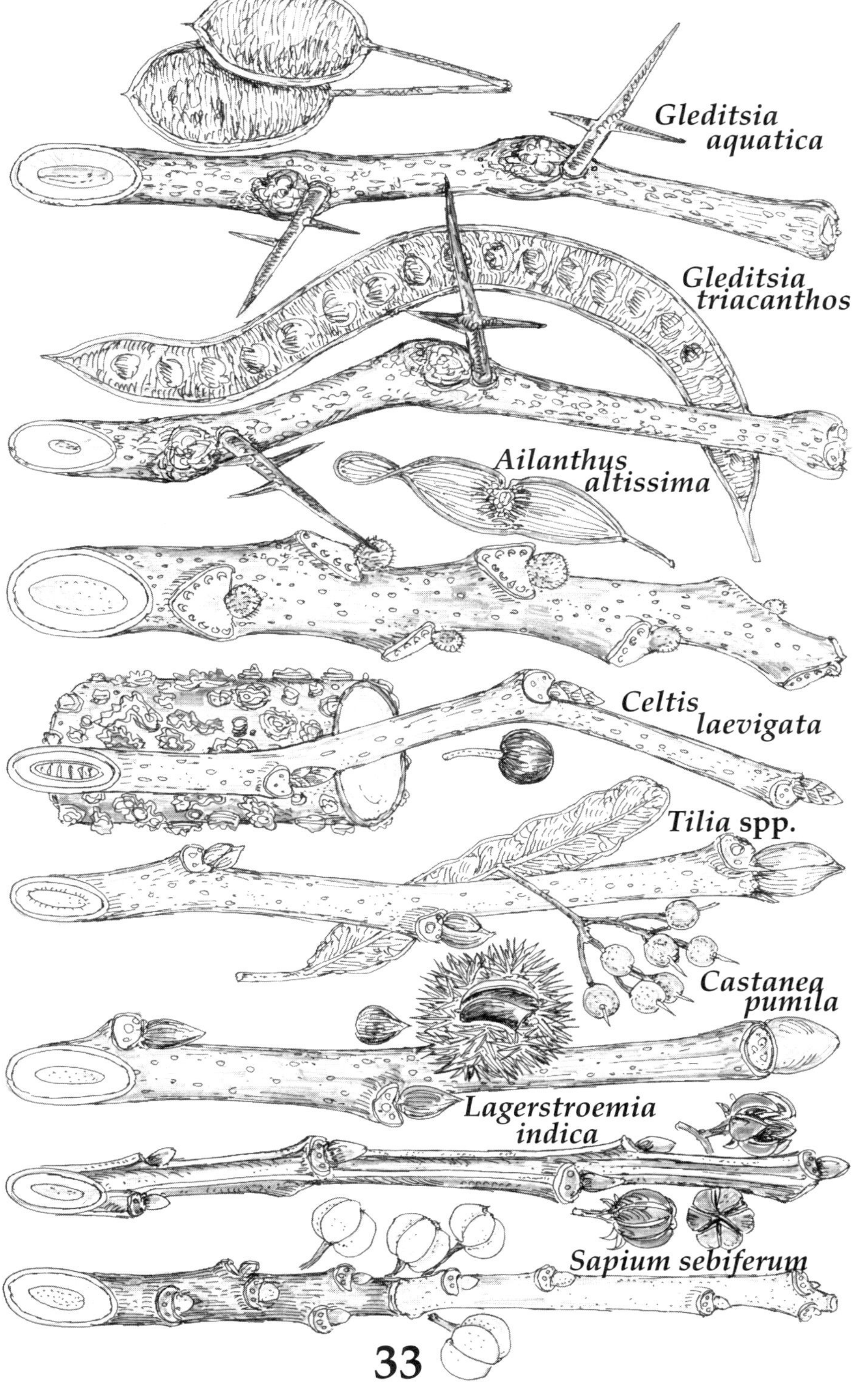
Gleditsia aquatica
Gleditsia triacanthos
Ailanthus altissima
Celtis laevigata
Tilia spp.
Castanea pumila
Lagerstroemia indica
Sapium sebiferum

37. Bundle scars numerous, often forming an ellipse; sap usually milky; breaking a twig reveals fine, silky white hairs across the break;
. . . . .red mulberry, *Morus rubra* ⇨
37. Bundle scars 3 or in 3 groups . . . . (38)

38. Bud scales with obvious vertical lines (striated), buds falcate-shaped; stipules may be persistent; bark of mature trees finely shaggy and easily peeled off in narrow strips;
. . . . eastern hophornbeam, *Ostrya virginiana* ⇨
38. Buds and bud scales not as above . . . . (39)

39. Buds flattened laterally and on both sides, giving the buds somewhat of a rectangular shape; main stems fluted with smooth, muscled appearance;
. . . . American hornbeam, *Carpinus caroliniana* ⇨
39. Buds not as above, bark not smooth or muscled . . . . (40)

40. Older twigs often with corky wings. . . (41)
40. Twigs without corky wings . . . . (42)

41. Buds sharp-pointed, obliquely sessile; accessory buds rarely present on top of leaf scar; twigs pubescent to glabrous; . . . . winged elm, *Ulmus alata* ⇨
41. Buds acute but not sharp-pointed; twigs slender and brown colored, pubescent to densely hairy; lateral buds obliquely sessile; a small accessory bud often present in a collateral position on top of leaf scar;
. . . . cedar elm, *Ulmus crassifolia* ⇨

42. Bark light gray-brown with large shreddy strips revealing orange-brown inner bark when removed; natural habitat is swampy areas, stream bottoms and overflow bottomland; buds somewhat globose;
. . . . planertree, *Planera aquatica* ⇨
42. Bark not as above . . . . (43)

43. Buds and twigs brown, not hairy, bud scales brown at base & dark brown on upper edges; cross-section into bark reveals alternating layers of buff and reddish-brown inner bark; #26
. . . . American elm, *Ulmus americana*. . . . . . ⇨
43. Buds dark, rusty brown to nearly black with red hairs at the top; twigs and buds usually hairy; cross-section into bark reveals inner bark reddish-brown only; . . . . .slippery elm, *Ulmus rubra*. . . . . . ⇨ #27

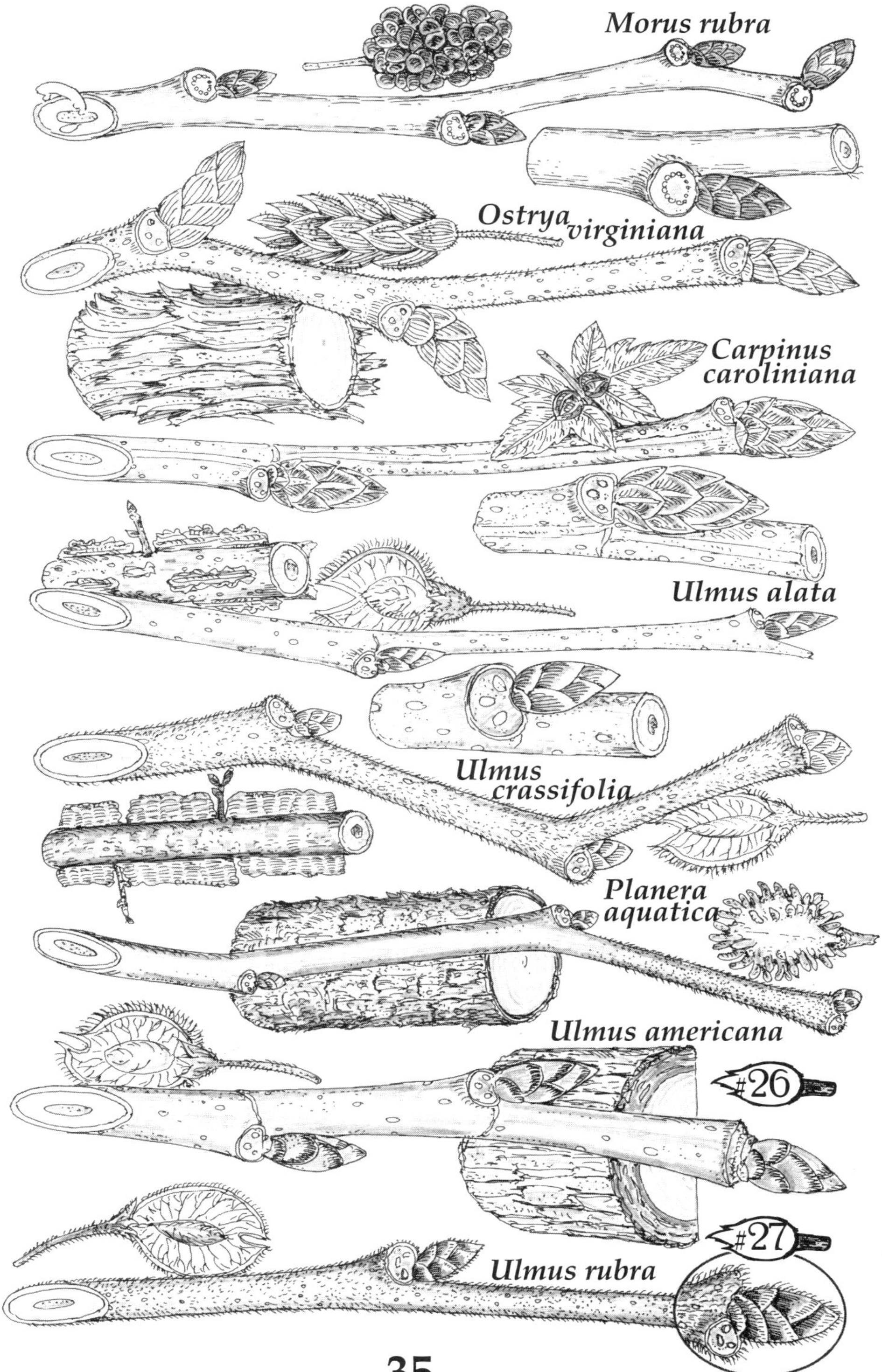
Morus rubra
Ostrya virginiana
Carpinus caroliniana
Ulmus alata
Ulmus crassifolia
Planera aquatica
Ulmus americana
#26
#27
Ulmus rubra

Key to *Carya* spp.
(hickories)

1. Terminal bud scales distinctly imbricate, usually more than 2 . . . .(2)
1. Terminal bud scales paired, valvate but often very pubescent and appearing naked. . . . .(6)

  2. Twigs light gray to mottled with patches of darker gray; buds golden-hairy; lateral buds indistinctly scaled, hairy and somewhat conical to oblong shaped; . . . . .nutmeg hickory; *Carya myristiciformis* ⇨
  2. Twigs brown to almost black. . . . .(3)

3. Outer scales of buds persistent, black; twigs red brown, usually glabrous to slightly pubescent; bark of mature stems shaggy and peeling off in large plates;. . . . . .shagbark hickory, *Carya ovata* ⇨
3. Outer bud scales soon deciduous, exposing lighter ones beneath; twigs often hairy and pleasantly fragrant when crushed; bark on mature stems never shaggy. . .(4)

  4. Terminal buds large, 10-25 mm; fruit husk greater than 6 mm thick; rachis very hairy; *(Carya alba)* . . . . mockernut hickory, *Carya tomentosa* ⇨
  4. Terminal buds usually 6-13 mm; fruit husk 6 mm thick or less. . . . . (5)

5. Buds glabrous or pale hairy; twigs glabrous; #28 . . . . .swamp hickory, *Carya leiodermis*. . . . .⇨
5. Buds and twigs reddish-brown, slightly pubescent; dry upland sites; . . . . black hickory, *Carya texana* ⇨

  6. Buds sulfur-yellow to golden-brown; . . . . .bitternut hickory, *Carya cordiformis* ⇨
  6. Buds brownish to gray-brown. . . . .(7)

7. Nuts twice as long as broad, smooth; kernel sweet tasting; . . . . pecan, *Carya illinoinensis* ⇨
7. Nuts as long as wide, usually 4-angled and very wrinkled, kernel very bitter tasting; saturated soils in bottomlands;. . . . . . water hickory, *Carya aquatica* ⇨

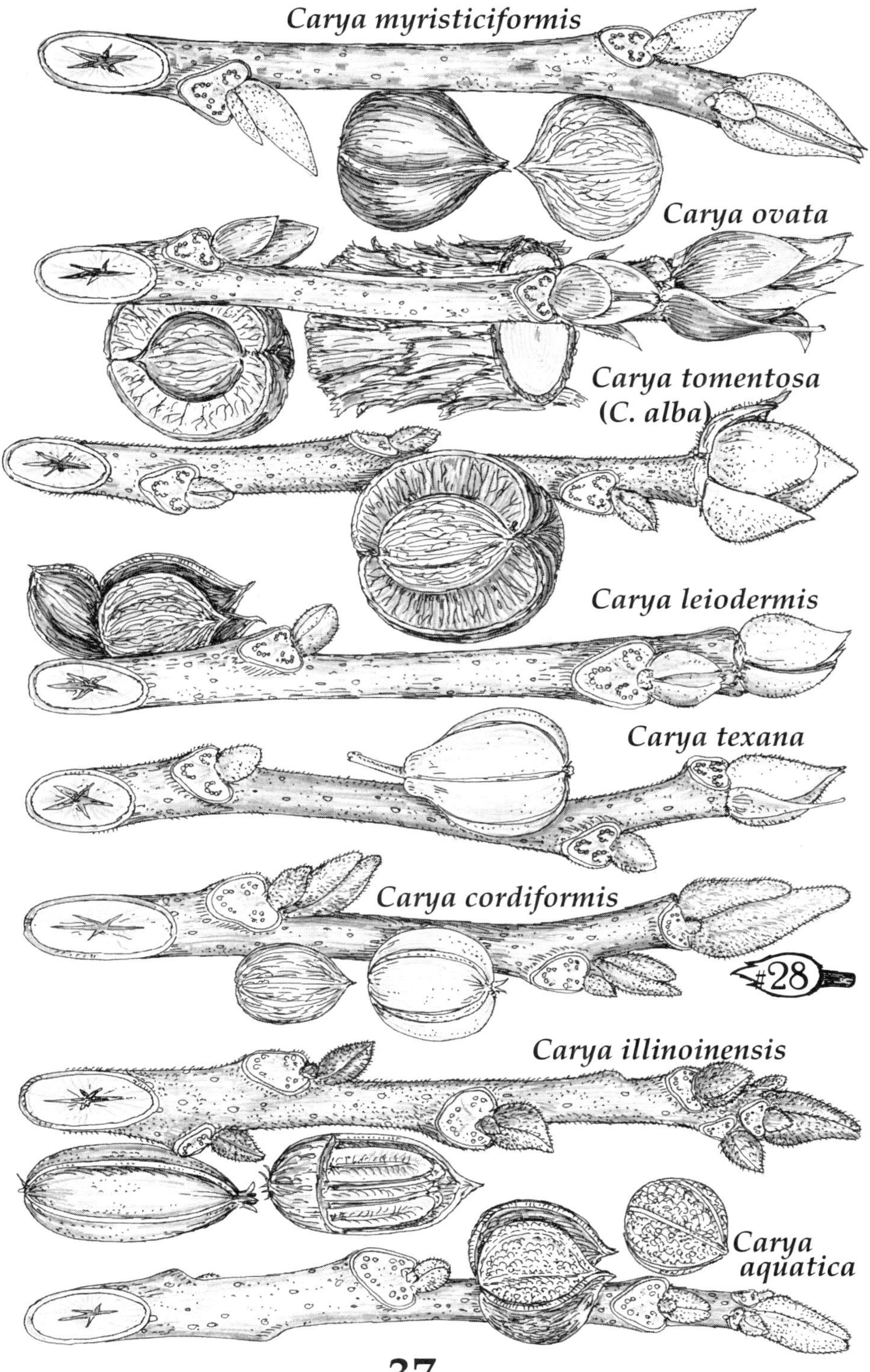
Carya myristiciformis
Carya ovata
Carya tomentosa
(C. alba)
Carya leiodermis
Carya texana
Carya cordiformis
#28
Carya illinoinensis
Carya
aquatica

Key to *Quercus* spp. (oaks)

1. **Leaves persistent (see an appropriate growing season key); buds 2 mm long, bud scales light brown with white scarious margins;. . . . . live oak, *Quercus virginiana* ⇨**
1. **Leaves mostly deciduous (occasionally tardily deciduous and often marcescent). . . . .(2)**

  2. **Largest terminal buds mostly 6-9 mm long, usually acute. . . . .(3)**
  2. **Largest terminal buds mostly less than 6 mm long, may be acute or obtuse. . . . . (7)**

3. **Buds mostly glabrous, dull light brown colored; edges of bud scales pale and scarious; top view of buds shows 5 sides; . . . . Shumard oak, *Quercus shumardii* ⇨**
3. **Buds pubescent to tomentose . . . .(4)**

  4. **Buds reddish, rusty brown, distinctly pubescent; . . southern red oak, *Quercus falcata*. . . . . ⇨** #29
  4. **Buds tomentose . . . .(5)**

5. **Buds gray tomentose; twigs often shiny; inner bark yellow-orange; . . . . black oak, *Quercus velutina* ⇨**
5. **Buds rusty tomentose; twigs not shiny . . . . .(6)**

  6. **Twigs distinctly ridged, not terete; lateral buds distinctly imbricately scaled; bark on mature trees shaggy; mesic lower slopes and well-drained small ridges in bottomlands; . . . . .swamp chestnut oak, *Quercus michauxii* ⇨**
  6. **Twigs indistinctly ridged; hairs on lateral buds make it somewhat difficult to distinguish imbricate scales; bark black and distinctly blocky; dry, sandy, upland sites; . . . . .blackjack oak, *Quercus marilandica* ⇨**

7. **Buds obtuse or nearly globose. . . . .(8)**
7. **Buds acute, relatively sharp-pointed . . . . . .(11)**

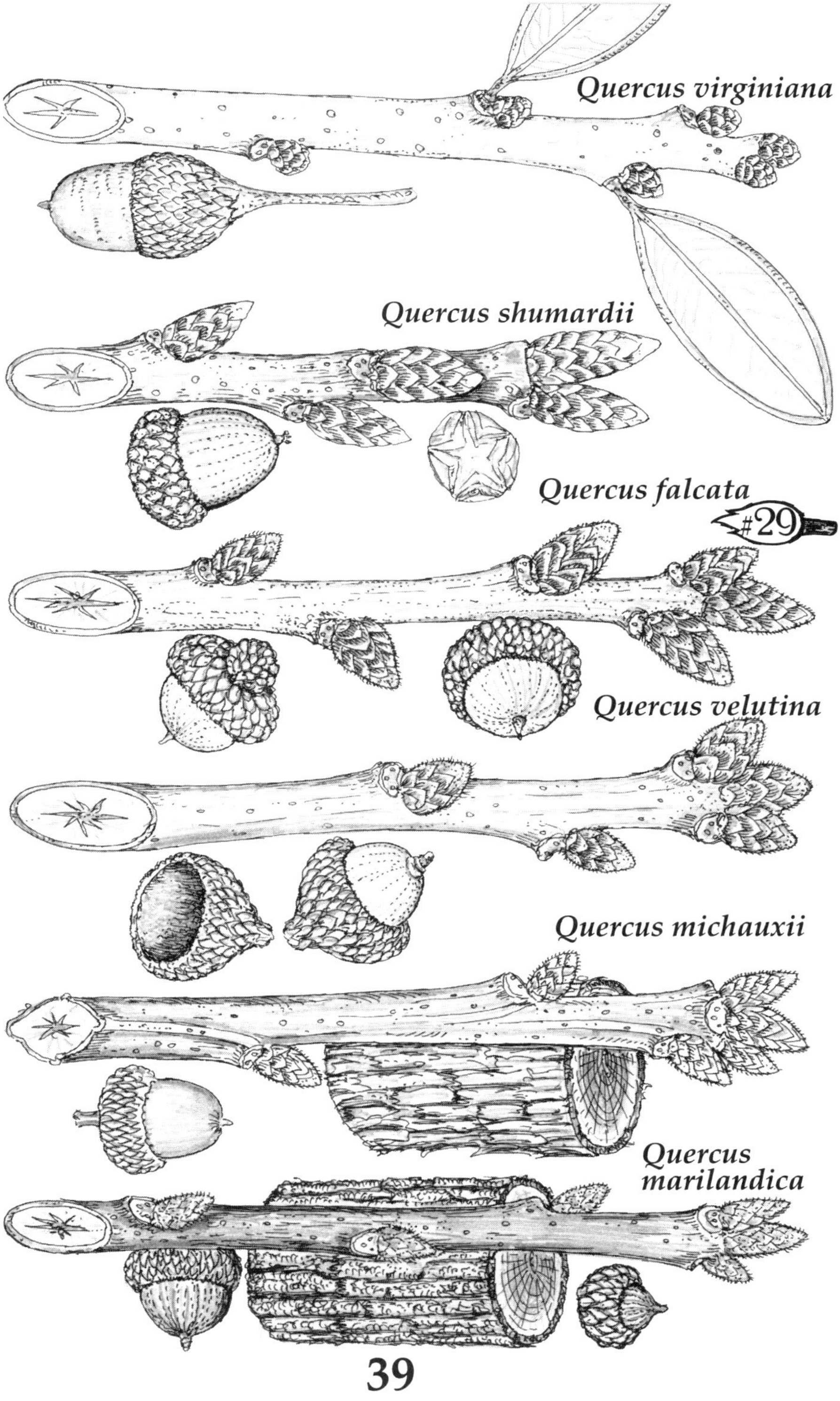
Quercus virginiana
Quercus shumardii
Quercus falcata
#29
Quercus velutina
Quercus michauxii
Quercus marilandica

8. Twigs relatively slender; twigs distinctly ridged near the tips; bark of mature stems very shaggy, peeling in broad plates on the edges; mesic slopes;
. . . . .white oak, *Quercus alba* ⇨

8. Twigs relatively stout; bark light gray-brown, may be loose but generally not shaggy. . . . . (9)

9. Twigs scurfy to tomentose; buds reddish-brown and distinctly 5-sided when viewed from the top; stipules not commonly persistent on terminal buds; upland sites; #30
. . . . post oak, *Quercus stellata*. . . . . . . ⇨

9. Twigs not scurfy, usually glabrous to slightly pubescent; stipules often persistent among top buds . . . .(10)

10. Buds gray pubescent, globose; acorn large with prominently fringed top edge
. . . . .bur oak, *Quercus macrocarpa* ⇨

10. Buds ovoid-globose, only slightly pubescent; twigs some what shiny, with numerous narrow greenish black cracks; saturated soils in river bottoms; acorn 13-25 mm long, 2/3 or nearly all of nut enclosed in the cup;
. . . . overcup oak, *Quercus lyrata* ⇨

11. Twigs densely pubescent to tomentose, gray to brown; leaves commonly marcescent. . . . . (12)

11. Twigs essentially glabrous or only slightly pubescent. . . . .(13)

12. Terminal buds distinctly pubescent; #29
. . . . .southern red oak, *Quercus falcata*. . . . . . . ⇨

12. Terminal buds only slightly pubescent, mostly glabrate; . . . . .cherrybark oak, *Quercus pagoda* ⇨

13. Largest terminal buds approximately 4-5 mm long . . . .(14)

13. Largest terminal buds 2-4 mm long. . . .(16)

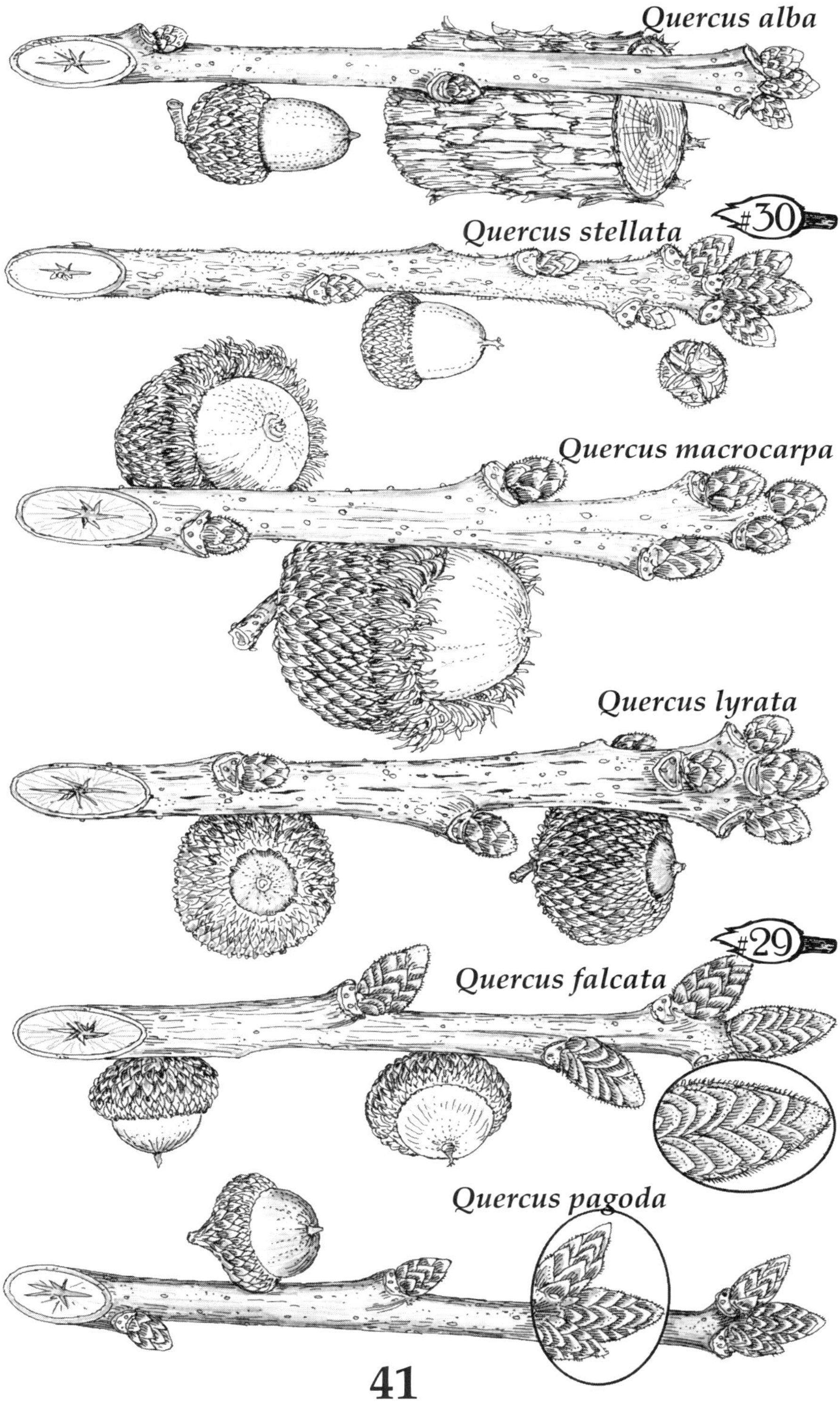
Quercus alba
#30
Quercus stellata
Quercus macrocarpa
Quercus lyrata
#29
Quercus falcata
Quercus pagoda

14. Buds scales dark gray brown, whitish-pubescent; twigs dark brown to gray brown; leaves often tardily deciduous;
. . . . water oak, *Quercus nigra* ⇨

14. Buds scales reddish-brown . . . .(15)

15. Twigs distinctly ridged, twigs light brown to tan brown; buds with white pubescence on top edges; mature stems shaggy with small plates;
. . . . .chinkapin oak, *Quercus muehlenbergii* ⇨

15. Twigs mostly terete, not ridged; buds with scarious margins, slight pubescence;
. . . . .willow oak, *Quercus phellos* ⇨

16. Typical habitat is dry, sandy uplands; bark distinctly blocky and gray-black;
. . . . bluejack oak, *Quercus incana* ⇨

16. Typical habitat is bottomlands or mesic slope positions. . . . .(17)

17. Mesic, well-drained sites; twigs generally reddish-brown to tan brown with distinct small ridges, especially near the tips; bud scales reddish; bark loose on mature stems, shaggy, peeling along edges; inner surface of acorn cup glabrous, kernel is sweet tasting, acorn about 20 mm long with 6 mm of its length enclosed in the cup; .white oak, *Quercus alba* ⇨

17. Saturated soils (true bottomland along rivers); twigs more terete, darker brown to gray-brown;
. . . . (18)

18. Leaves often sub-persistent or tardily deciduous;
. . . . .swamp laurel oak, *Quercus laurifolia* ⇨

18. Leaves not tardily deciduous;
. . . . Nuttall oak, *Quercus texana* ⇨

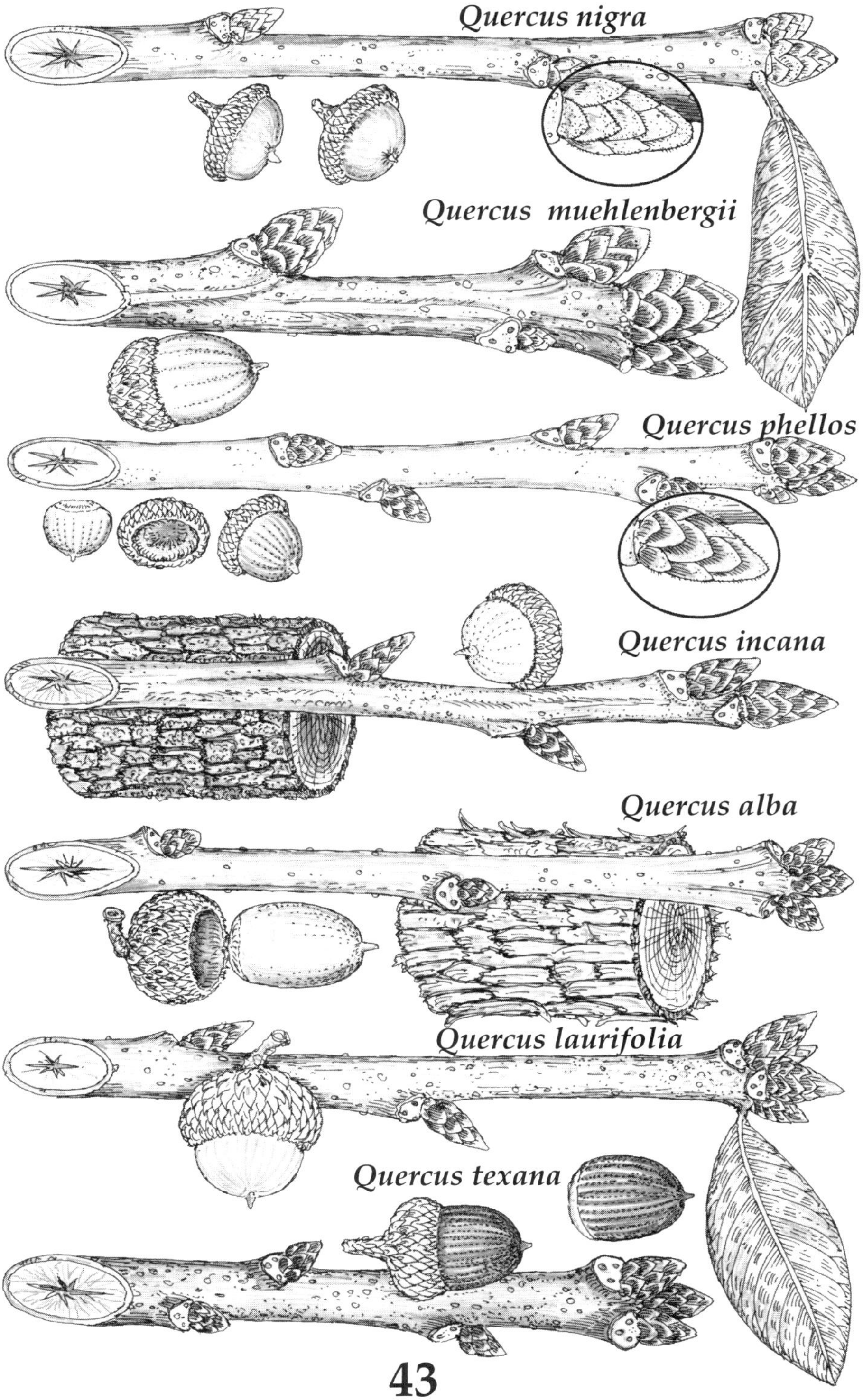
Quercus nigra
Quercus muehlenbergii
Quercus phellos
Quercus incana
Quercus alba
Quercus laurifolia
Quercus texana

ILLUSTRATED GLOSSARY

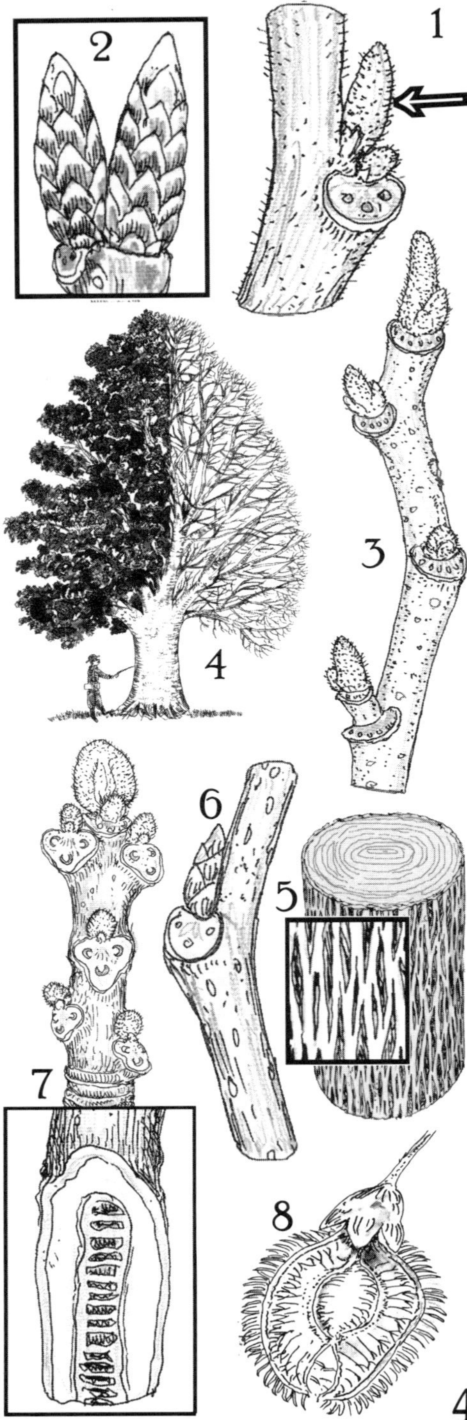

accessory bud: An additional or extra bud other than the lateral bud located above the leaf scar. (1. *Hamamelis virginiana*)

acute: Tapering to a point. (2. *Quercus shumardii* buds)

alternate: Occurring one at each node, scattered singly and spirally along the stem.
(3. *Toxicodendron radicans*)

arborescent: Tree-like in appearance.
(4. *Fagus grandifolia*)

anastomosing: Interlacing, generally referring to a bark pattern where furrows and ridges are interlacing.
(5. *Fraxinus americana* bark)

appressed: Flattened against. (6. *Celtis laevigata* bud)

chambered: Pith type that consists of small side-by-side chambers. (7. *Juglans nigra* twig)

ciliate: Margin fringed with hairs. (8. *Ulmus americana* samara fruit)

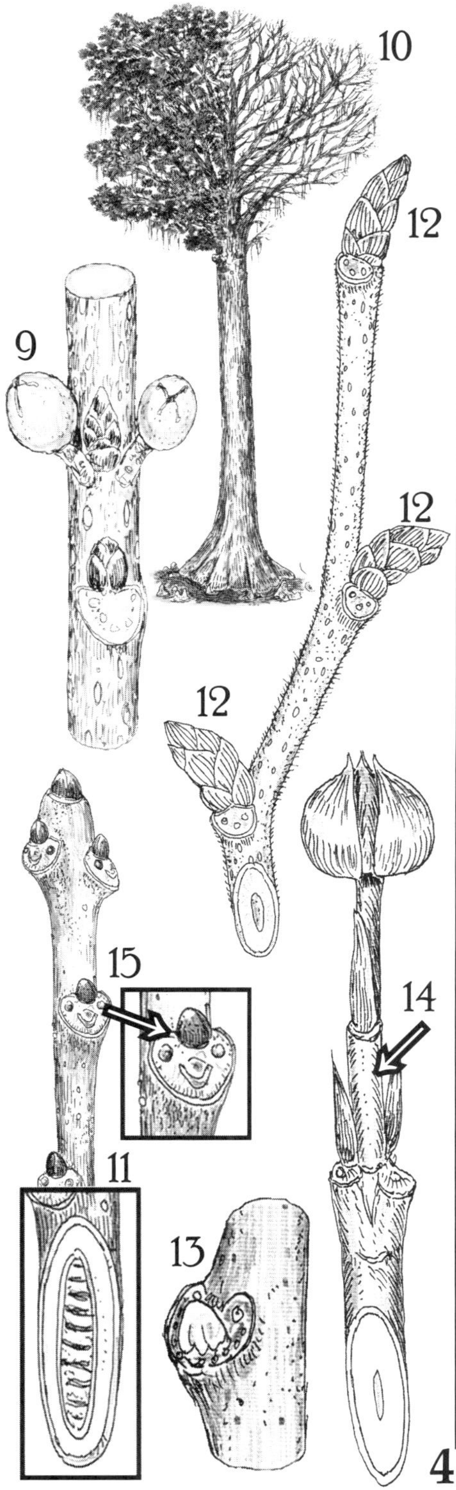

collateral buds: More than one bud located on top of leaf scar and they are arranged horizontally across the top of leaf scar. (9. *Lindera benzoin*)

deciduous: Leaves drop during the autumn.
(10. *Taxodium distichum*)

diaphragmed: Pith type that is solid (continuous) but has noticeable lines that divide the pith into segments.
(11. *Nyssa aquatica* twig)

falcate: Sickle- or scythe-shaped, curved.
(12. *Ostrya virginiana* buds)

glabrate: Becoming glabrous or nearly glabrous.

glabrous: Smooth, not hairy or rough to the touch.
(13. *Rhus glabra* twig)

glaucous: With a whitish, waxy covering.
(14. *Cornus florida* twig)

globose: Round or spherical in shape.
(15. *Nyssa aquatica* bud)

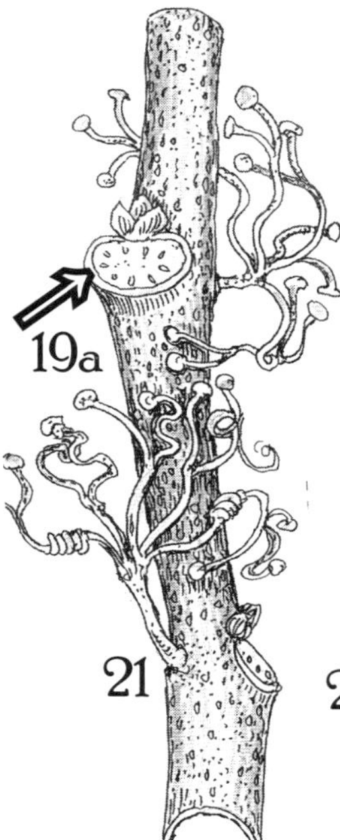

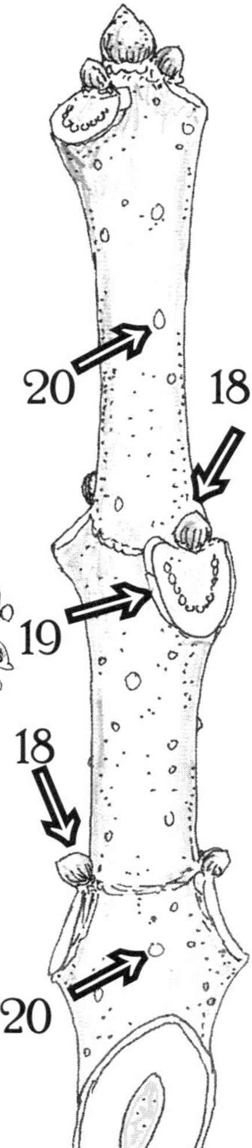

habit: Typical form a particular species usually maintains; also refers to the type of woody plant (tree versus shrub versus liana).

habitat: Typical environment where a species is usually found. (16. *Taxodium distichum* and *Nelumbo lutea*)

homogenous: Same or consistent.

imbricate: Overlapping, in shingle-like fashion. (17. *Liquidambar styraciflua* bud scales)

lateral bud: Bud found on sides of twig at each node, also called axillary bud. (18. *Fraxinus caroliniana*)

leaf scar: Scar remaining on the twig where the leaf had been attached. (19. *Fraxinus caroliniana* & 19a. *Parthenocissus quinquefolia*)

lenticel: Corky line or spot on the twig surface. (20. *Fraxinus caroliniana*)

liana: Woody vine. (21. *Parthenocissus quinquefolia*)

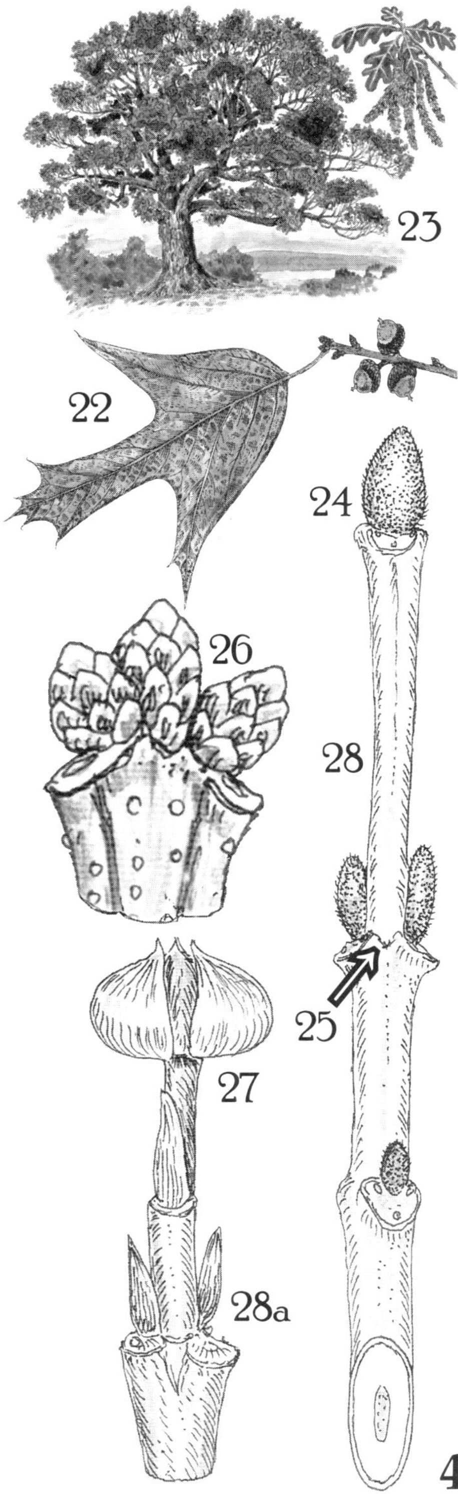

marcescent: Leaves wither, turn brown but persist on the twig until the next spring. (22. *Quercus falcata*)

mesic: Moisture regime intermediate between dry (xeric) and wet (hydric). (23. *Quercus alba*)

naked: Buds without scales; often very hairy. (24. *Viburnum rufidulum*)

node: Location on the twig where leaf scars are found. (25. *Viburnum rufidulum*)

obtuse: Blunt or rounded; angle greater than 90 degrees. (26. *Quercus alba* buds)

ovoid: Rounded. (27. *Cornus florida* flowering bud)

opposite: Two leaf scars at same node but on opposite sides. (28a. *Cornus florida* and 28. *Viburnum rufidulum*)

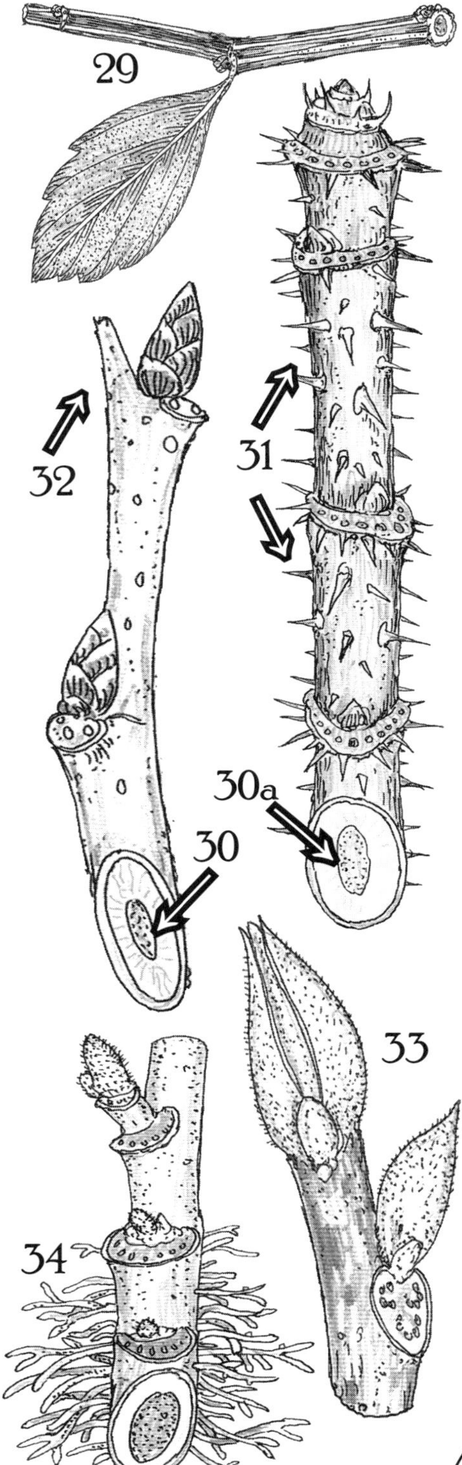

persistent: Remaining throughout most of winter. (29. *Baccharis halimifolia*)

pith: Central portion of twig, usually of different texture and color from rest of twig. (30a. *Aralia spinosa* and 30. *Ulmus alata*)

prickle: Arming that arises from epidermal cells along twig or bole. (31. *Aralia spinosa*)

pseudo-terminal bud: Same as false terminal bud, a lateral bud that has assumed a terminal position on a twig due to end of twig dying back; presence of branch scar indicates if bud on end of twig is a true or pseudo-terminal bud; often found on distinctly zigzag twigs. (32. *Ulmus alata*)

pubescence: Soft hairs evident on the structure. (33. *Carya myristiciformis* bud)

riparian: Associated with bodies of water, particularly moving streams. (e.g. *Betula nigra*)

rootlets: Root-like structures on the main stem of vines that adhere the vine to host plant. (34. *Toxicodendron radicans*)

35
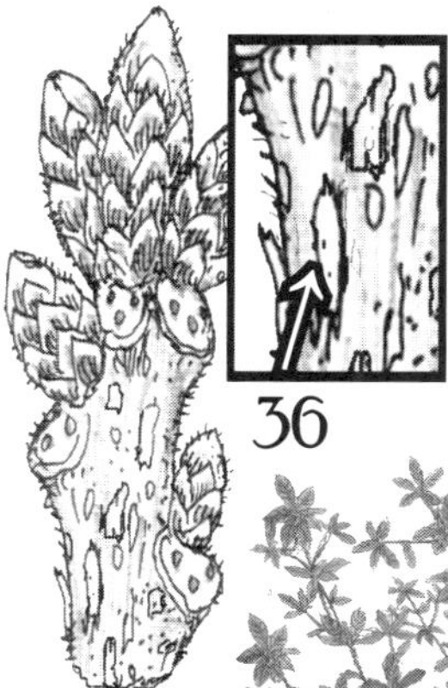
36

37
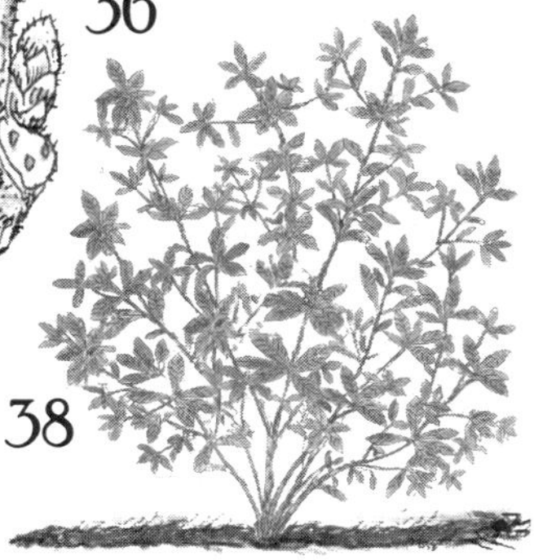
38
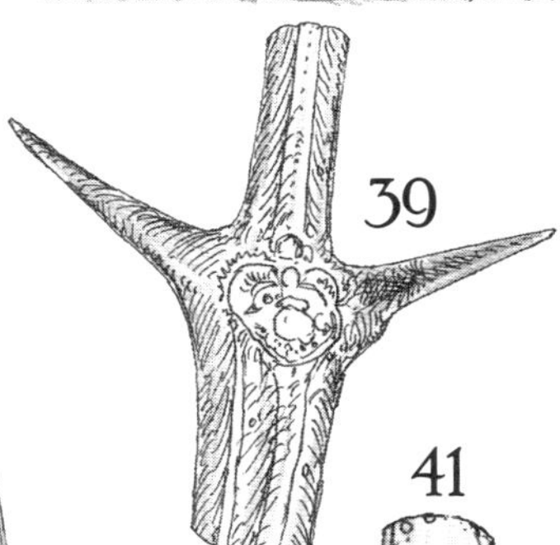
39
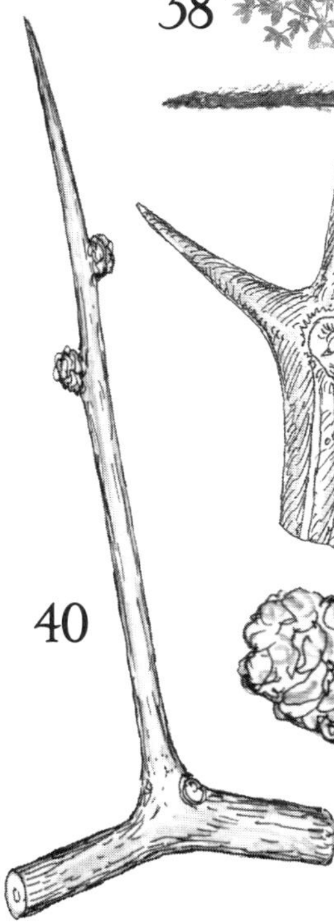
40

41

scarious: Margin of structure not smooth; irregular or jagged. (35. *Quercus shumardii* bud scales)

scurfy: Covered with small scales.
(36. *Quercus stellata* twig)

sessile: Without a stalk. (37. *Ulmus crassifolia* lateral bud)

shrub: Woody plant that at maturity in an optimum environment is generally less than 10-20 feet in height; often multi-stemmed from the base with stems upright, curved or lying prostrate on the ground. (38. *Rhododendron oblongifolium*)

spine: Hard, sharp-pointed structure derived from below the surface.
(39. *Robina pseudoacacia*)

spinescent twig: Spiny or spine-tipped; refers to short branches that terminate in a spine.
(40. *Sideroxlyon lanuginosum*)

spur branch: Indeterminate short shoot; short side branch; often appears knobby.
(41. *Maclura pomifera*)

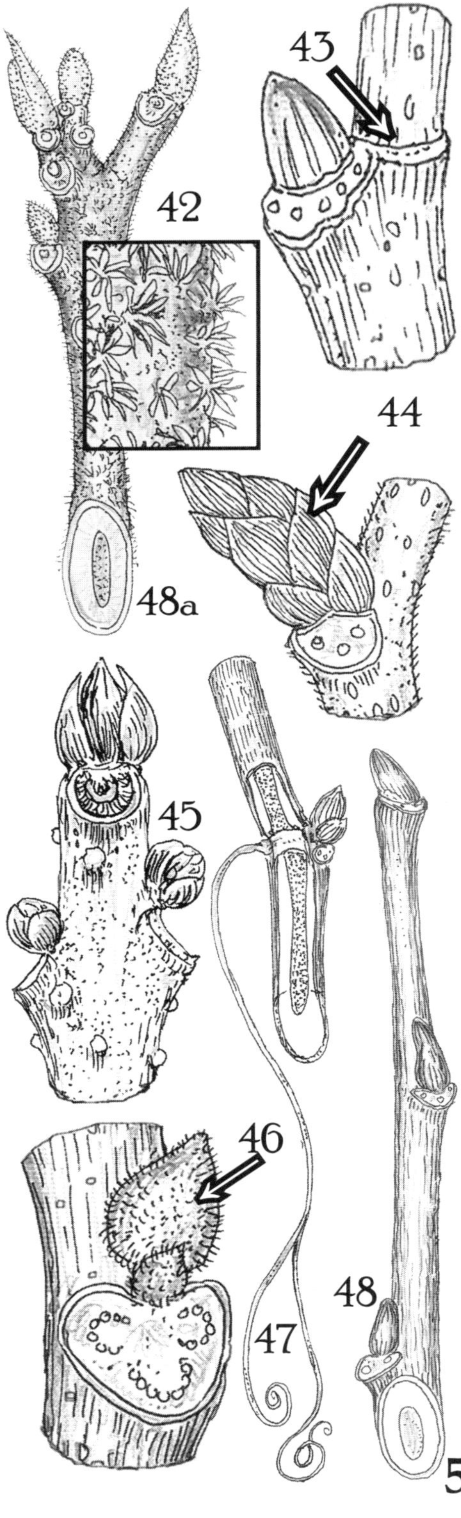

stellate: Hairs that are branched, often somewhat star-shaped.
(42. *Clethra alnifolia* twig)

stipule scar: Scars that occur after stipules fall off; found along both sides of the top of leaf scars (slit-like in appearance) or occurring as ring that encircles or nearly encircles the twig at the node.
(43. *Platanus occidentalis*)

striated: Lines or grooves that run vertically.
(44. *Ostrya virginiana* buds)

sub-opposite: Two leaf scars at each node but are slightly offset from each other.
(45. *Chionanthus virginicus*)

superposed bud: Accessory bud occurring at node above the lateral bud; more than one bud per node and they are arranged vertically above the leaf scar.
(46. *Carya illinoinensis*)

tendrils: Modified stems or leaves on vines which coil around host structures to provide support for the vine. (47. *Vitis aestivalis*)

terete: Circular (round) in cross-section.
(48. *Salix nigra* twig and 48a. *Clethra alnifolia* twig)

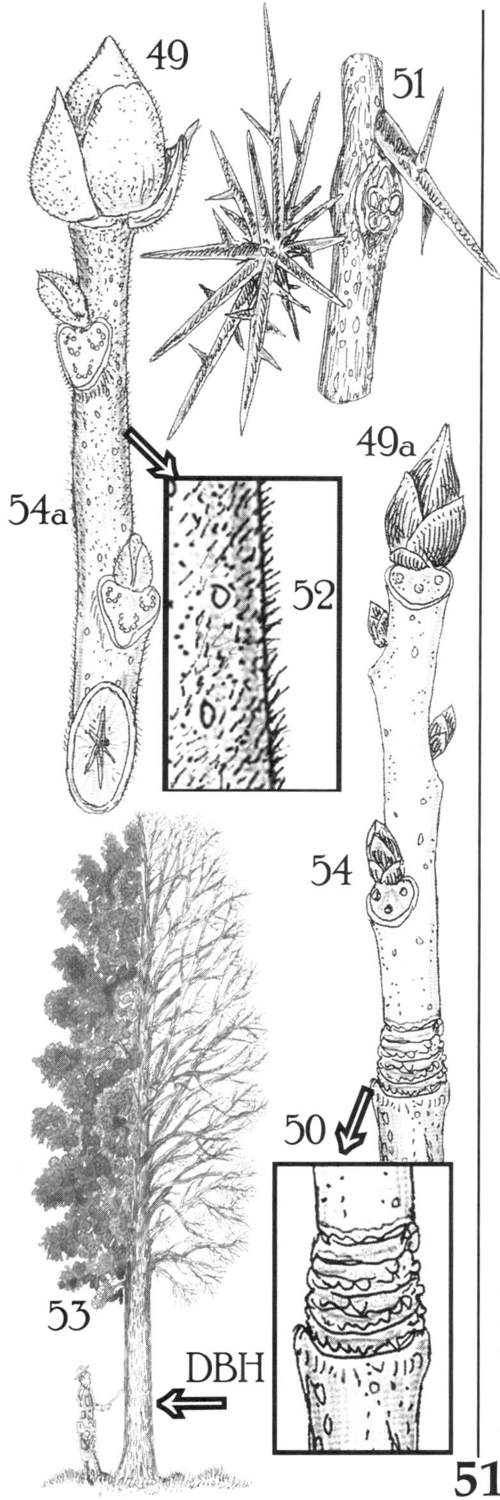

true terminal bud:

Bud which is at the end of a twig. (49. *Carya tomentosa* and 49a. *Nyssa sylvatica*)

terminal bud scale scars:

Ring of scars left from base of last year's terminal bud; marks the beginning of the current twig's growth. (50. *Nyssa sylvatica* twig)

thorn:

Sharp-pointed hard woody branches, positioned above the leaf scar or may terminate a twig (spinescent twig). (51. *Gleditsia aquatica*)

tomentose:

Having curled, woolly hairs. (52. *Carya tomentosa* twig)

tree:

Woody plant, that at maturity in an optimum environment, exceeds 10-20 feet in height, commonly has a single stem arising from the ground which exceeds 3 inches at DBH (4.5 feet above base of stem on uphill side of tree). (53. *Liriodendron tulipifera*)

twig:

Terminal or most recent growth of a branch; delimited commonly by the presence of terminal bud scale scars. (54. *Nyssa sylvatica twig* and 54a. *Carya tomentosa* twig)

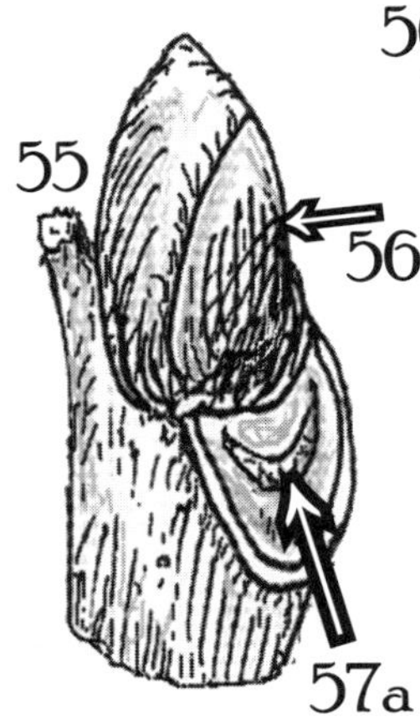

twig scar: Occurs on species that have pseudoterminal buds; it is the remnant or scar left from the dieback of the twig to the lateral bud that assumes the terminal position.
(55. *Diospyros virginiana*)

valvate: Refers to buds that are covered by 2-3 scales that do not overlap.
(56a. *Liriodendron tulipifera* & 56. *Diospyros virginiana*)

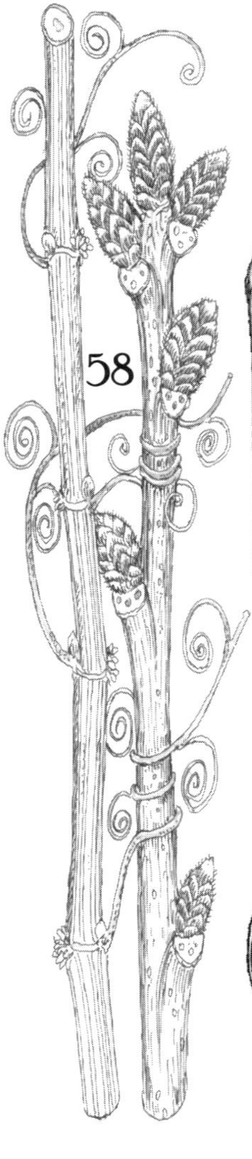

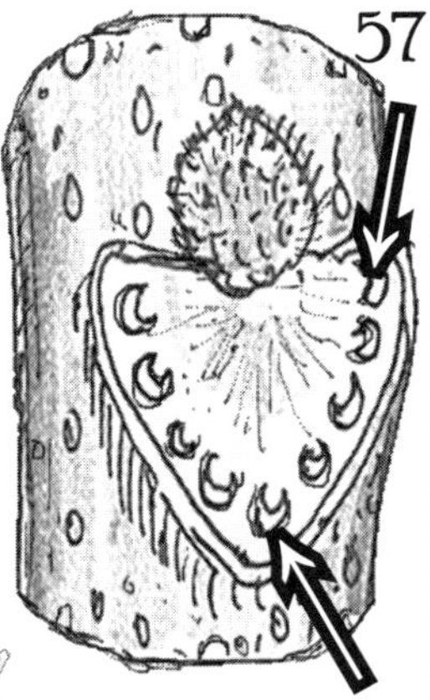

vascular bundle scar: Small dots or corky projections or slits that occur within leaf scars that show where the strands of vascular tissue passed from the twig into the leaf.
(57a. *Diospyros virginiana* & 57. *Ailanthus altissima*)

vine: Woody plant that at maturity in an optimum environment climbs upon a host structure by means of tendrils, aerial roots or by simply twining (coiling) around the host.
(58. *Brunnichia cirrhosa* on *Quercus* spp.)

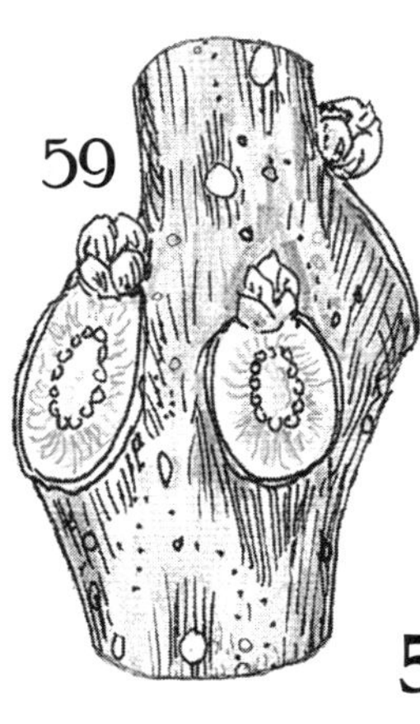

whorled: More than two leaf scars per node; often one scar may be slightly higher than the other two.
(59. *Catalpa speciosa*)

ILLUSTRATED TWIG GROWTH

Carya illinoinensis

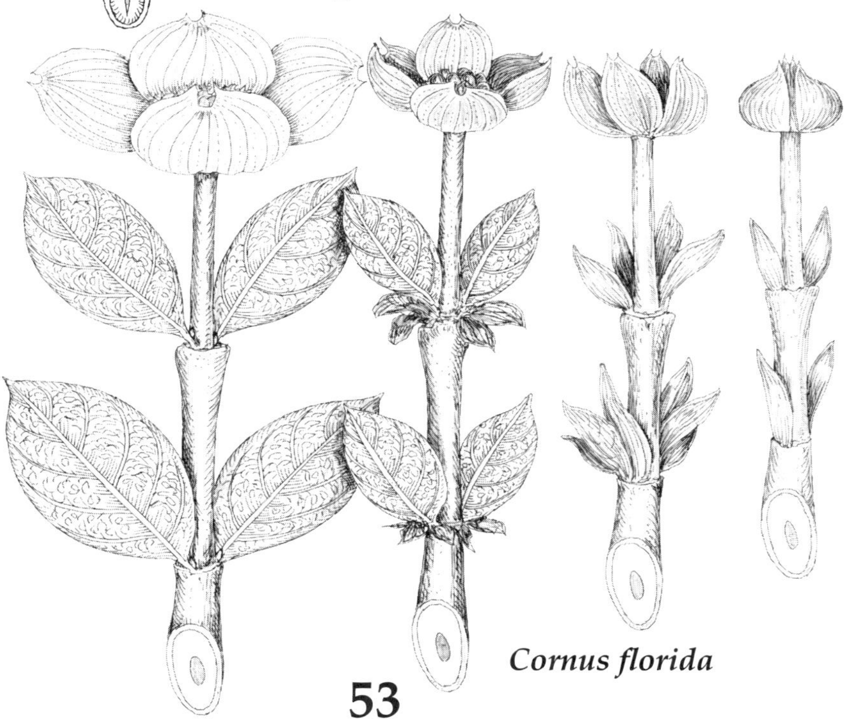

Cornus florida

ILLUSTRATED TWIG

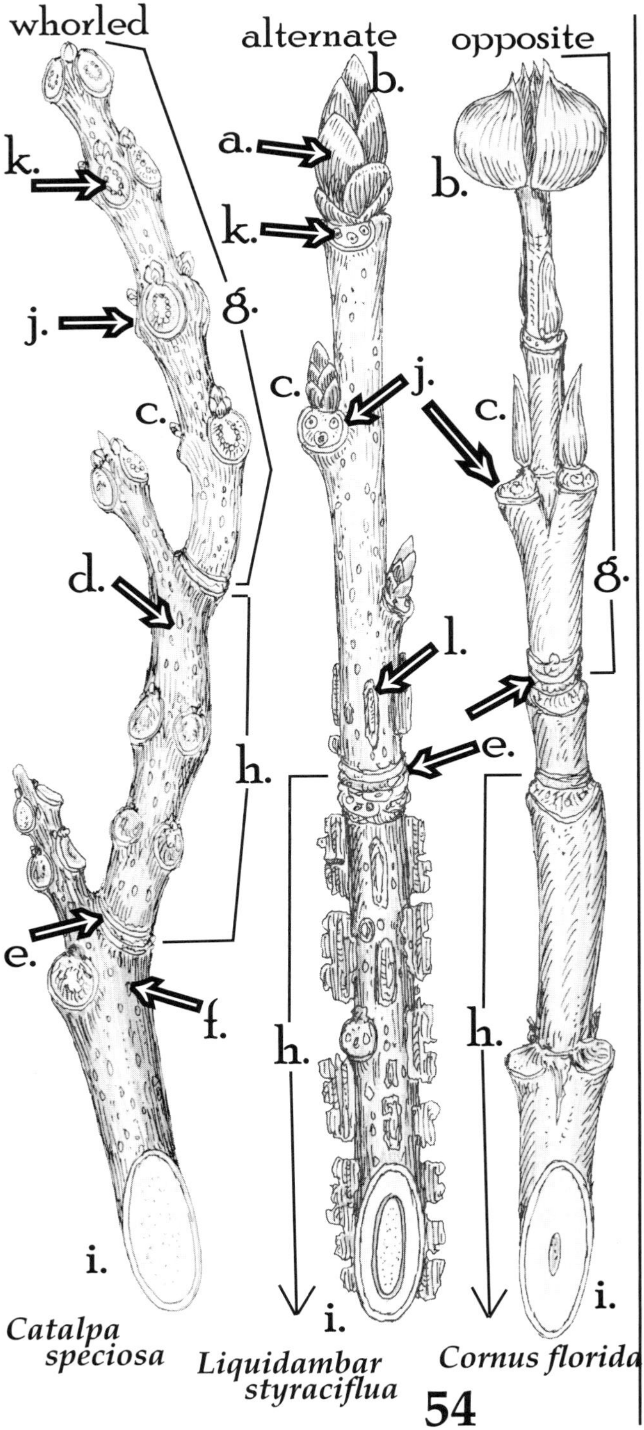

a. bud scale

b. terminal bud

c. axillary bud

d. lenticel

e. bud scale scars

f. node

g. first year's growth

h. second year's growth

i. pith

j. leaf scar

k. bundle scar

l. cork wings

CHARACTERISTICS

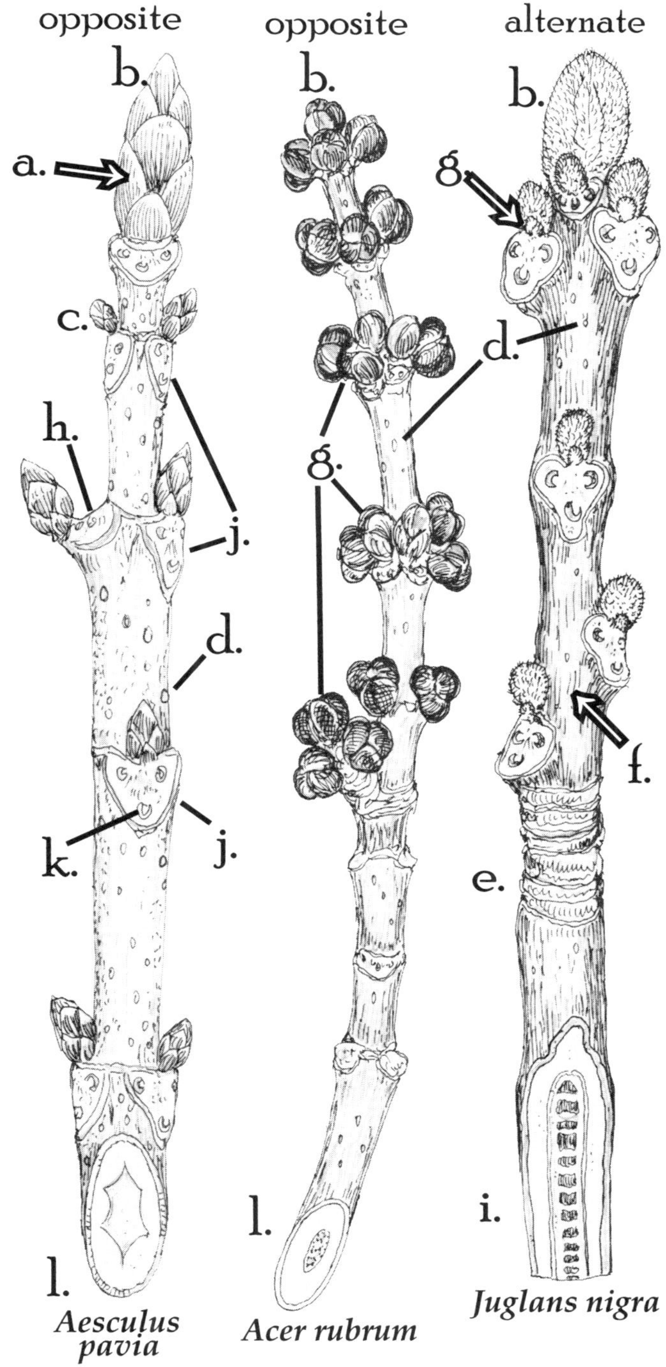

Aesculus pavia

Acer rubrum

Juglans nigra

a. bud scale

b. terminal bud

c. axillary bud

d. lenticel

e. bud scale scars

f. node

g. accessory buds

h. fruit scar

i. chambered pith

j. leaf scar

k. bundle scar

l. pith

ILLUSTRATED PITHS

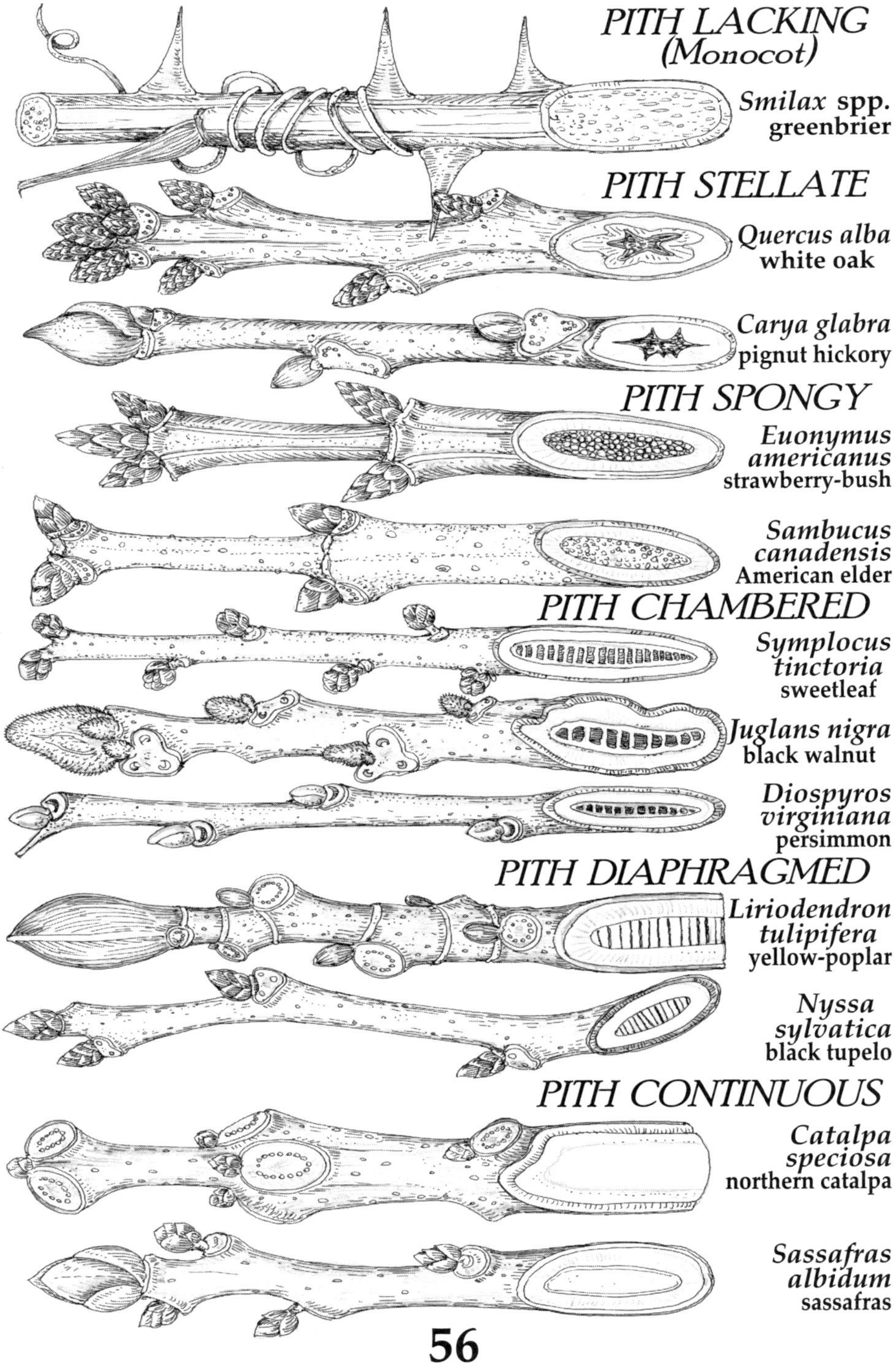

OTHER TREE ILLUSTRATIONS

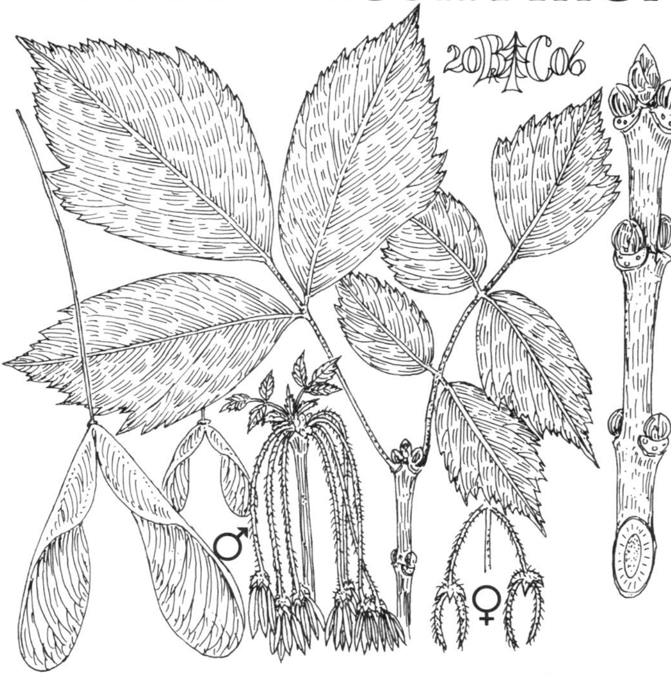

Acer negundo L.- boxelder

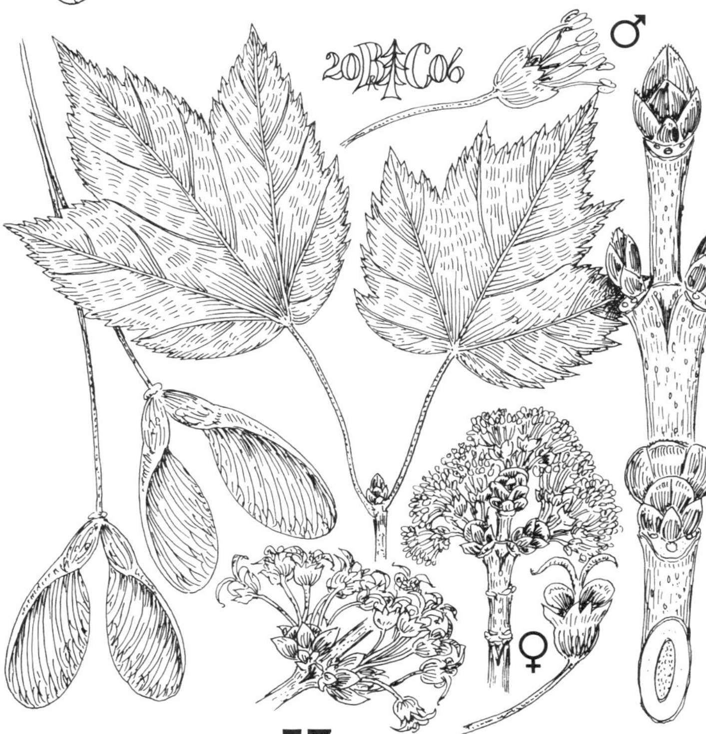

Acer rubrum L.- red maple

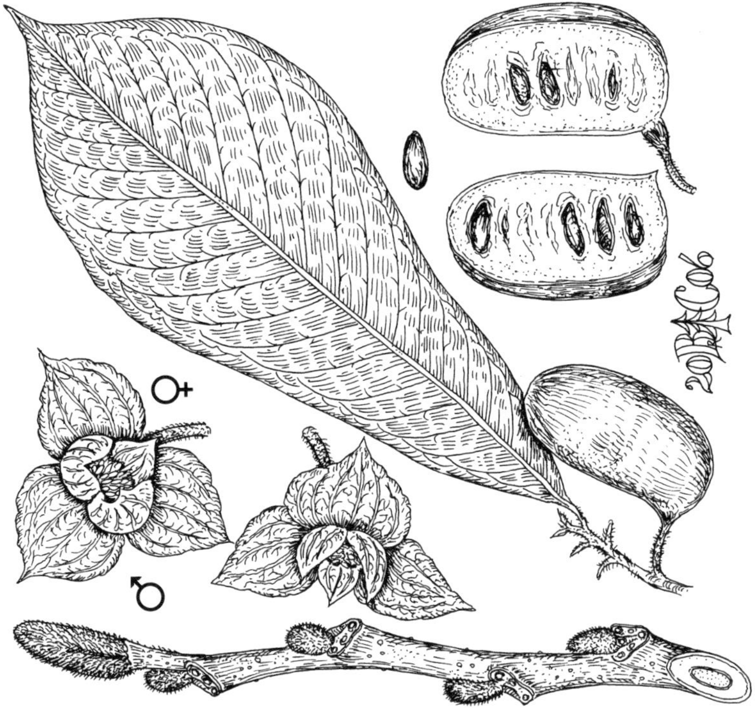

Asimina triloba (L.) Dunal - pawpaw

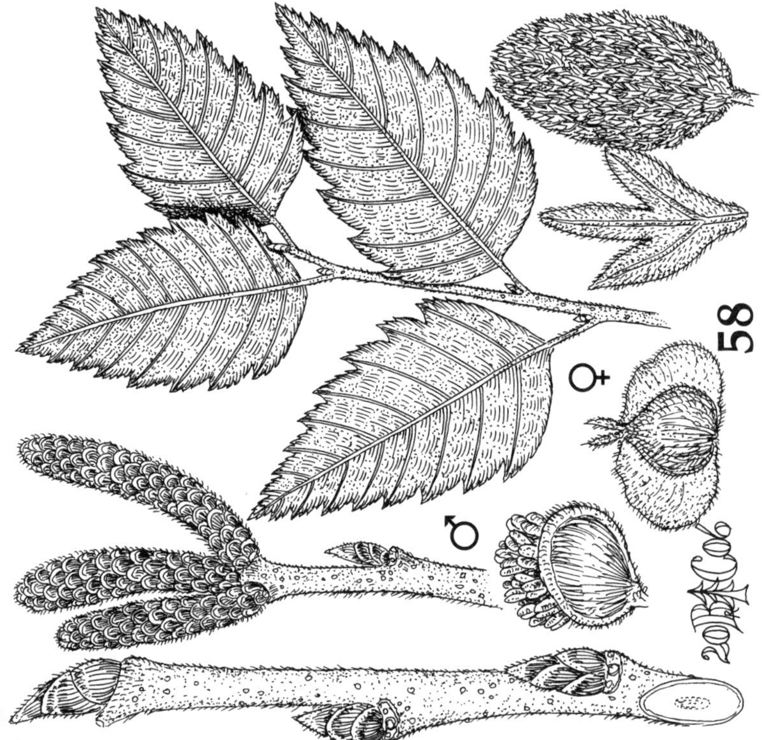

Betula nigra L. - river birch

Carpinus caroliniana Walt. - American hornbeam

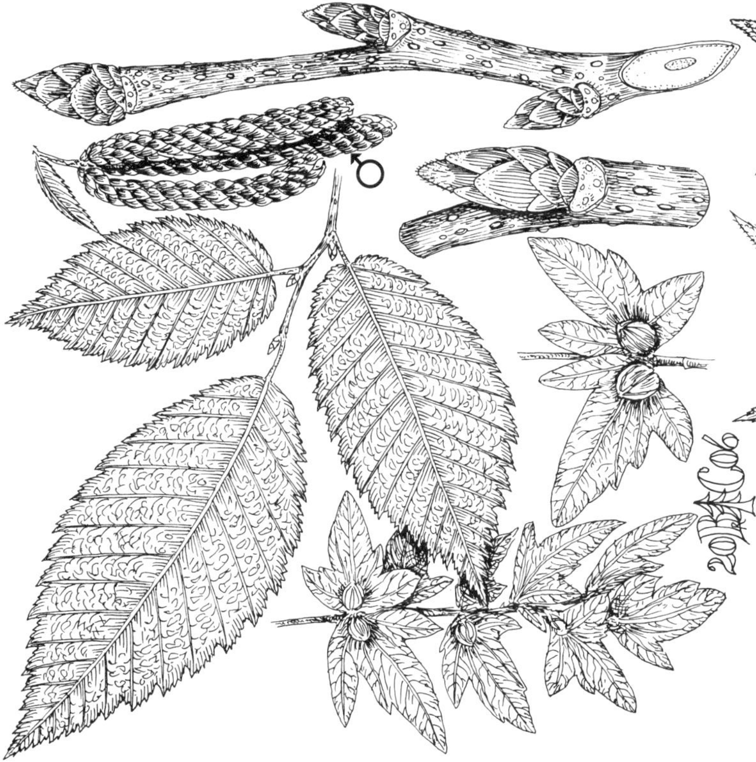

Carya cordiformis (Wangenh.) K. Koch - bitternut hickory

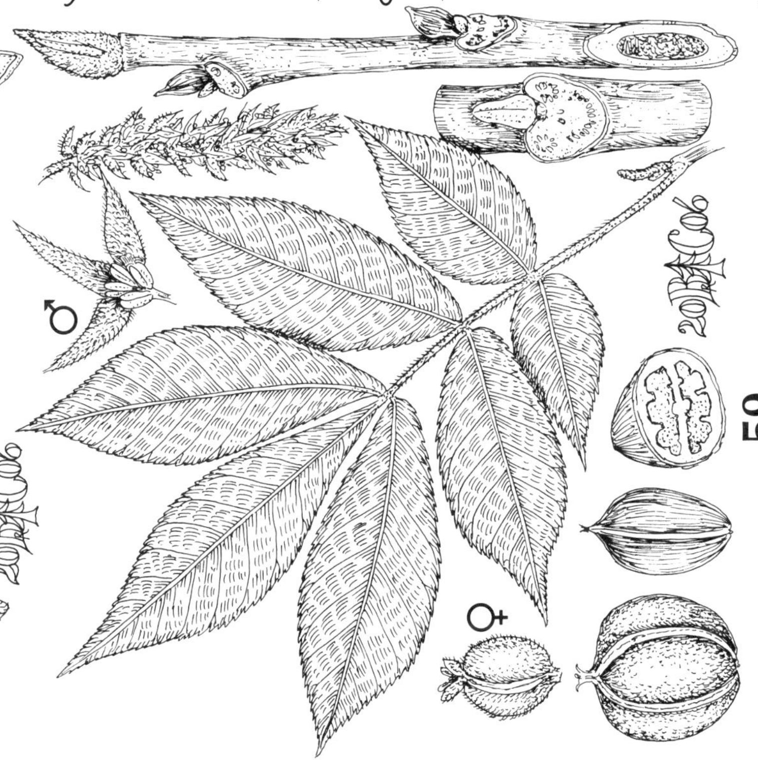

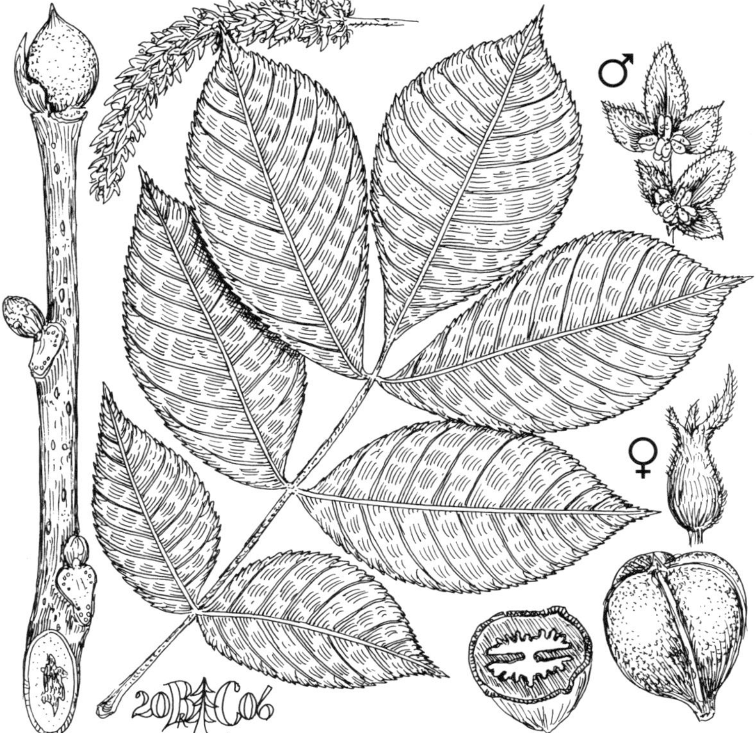

Carya glabra (Mill.) Sweet – pignut hickory

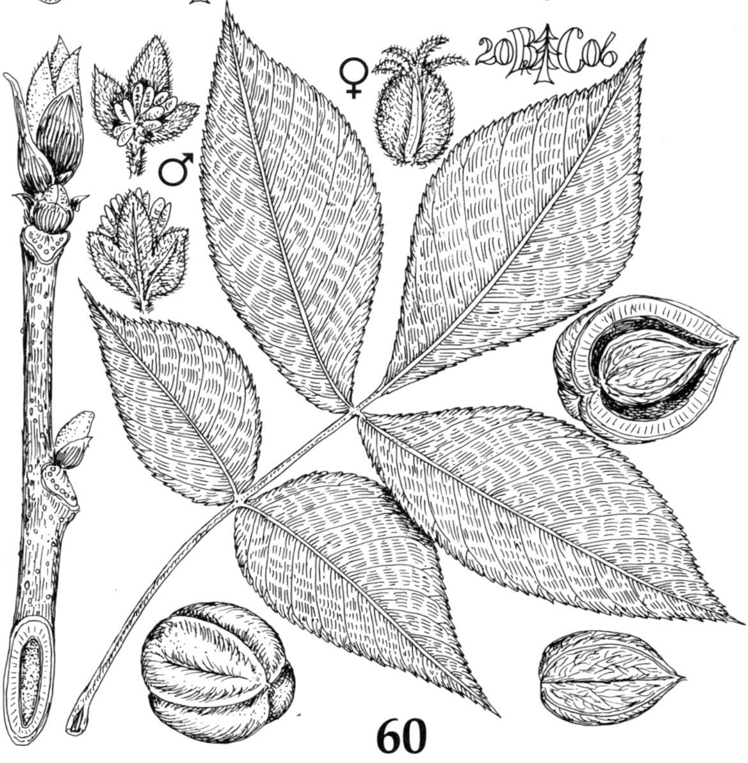

Carya ovata (P. Mill.) K. Koch – shagbark hickory

Carya tomentosa (Poir. ex Lam.) Nutt. - mockernut hickory
(*C. alba*)

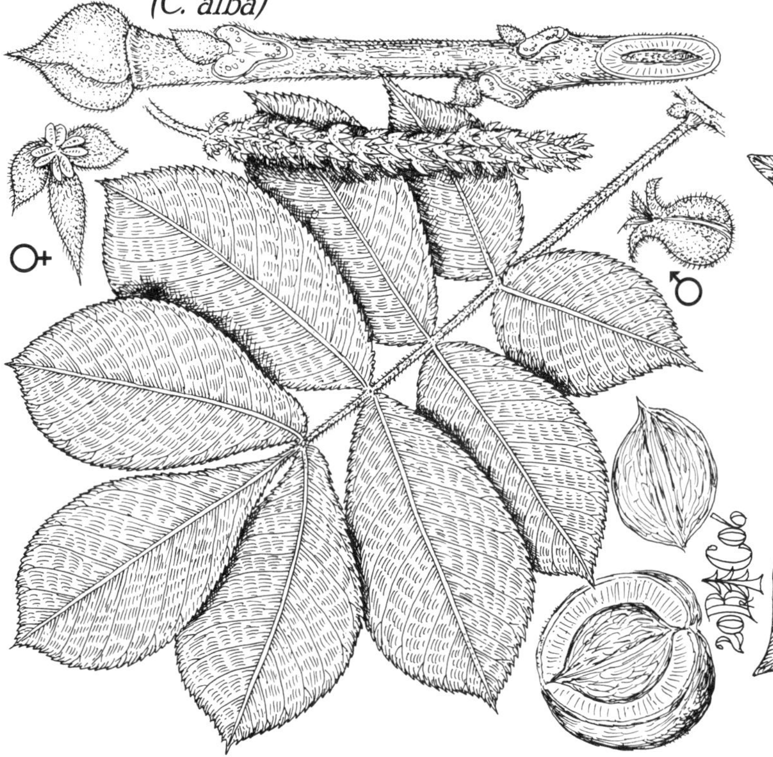

Cercis canadensis L. - eastern redbud

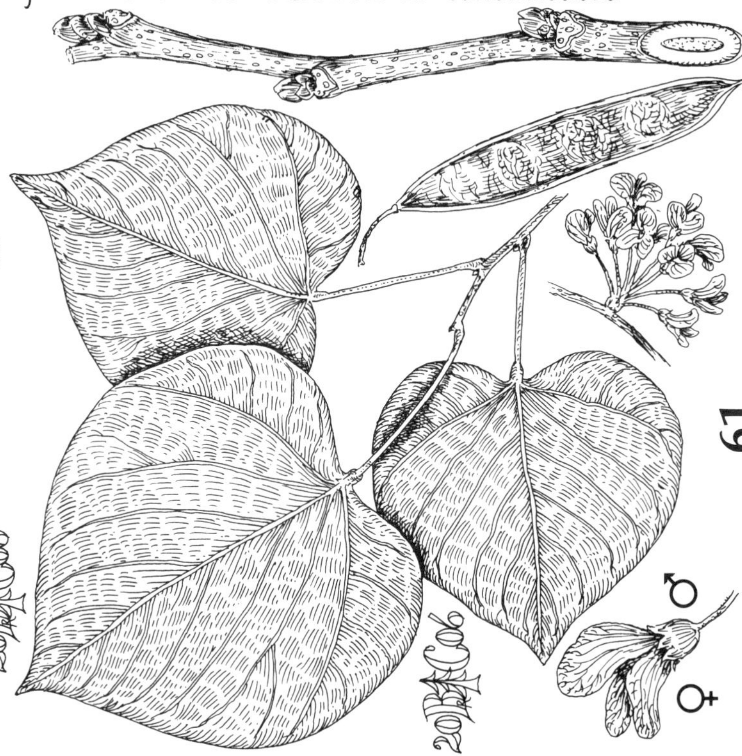

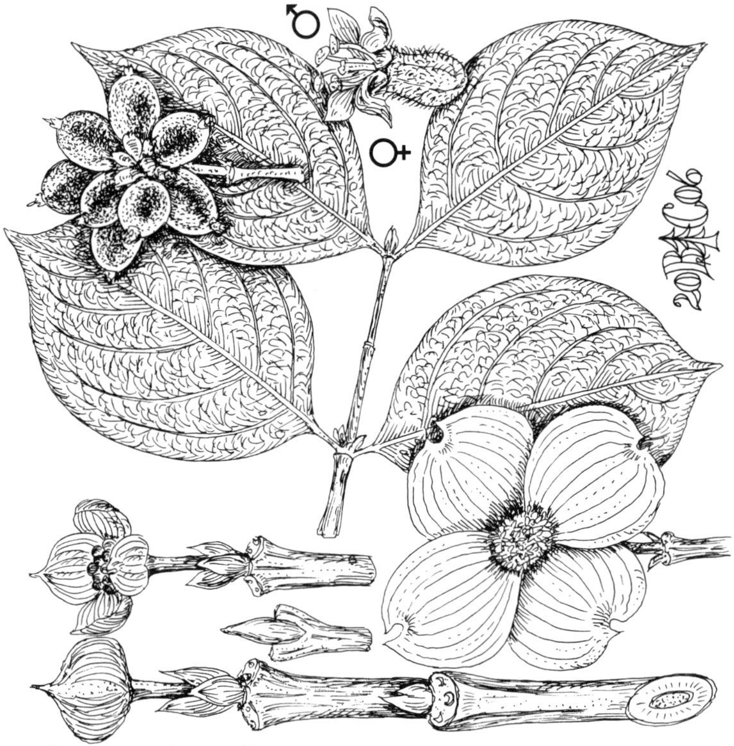

Cornus florida L.- flowering dogwood

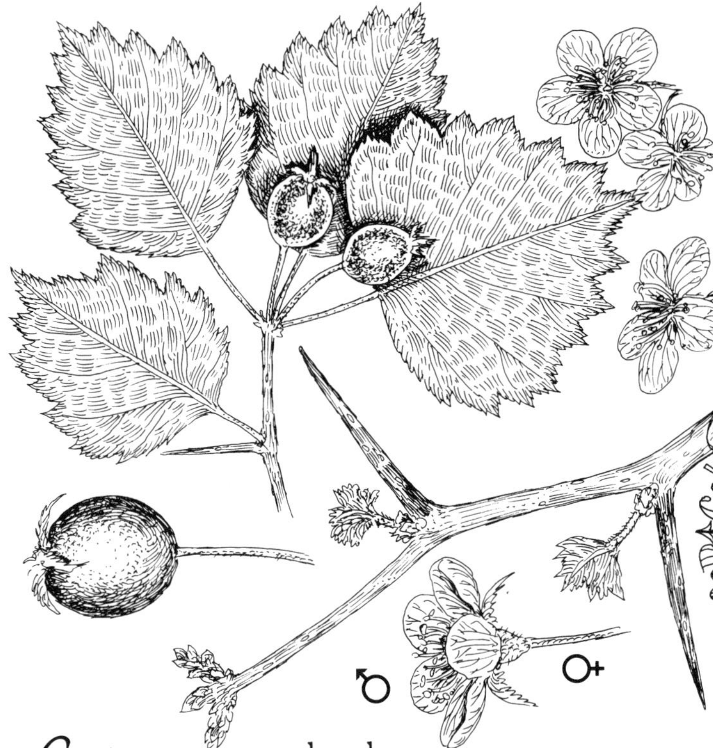

Crataegus spp.- hawthorns

Diospyros virginiana L. -common persimmon

Fagus grandifolia Ehrh. - American beech

Fraxinus americana L.-white ash

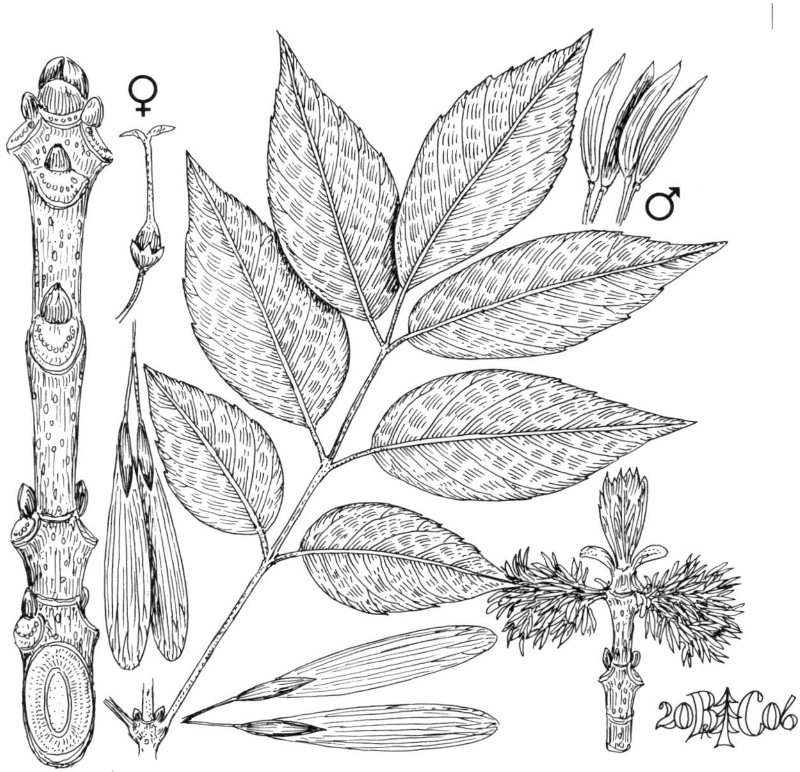

Gleditsia triacanthos L.-honeylocust

Hamamelis virginiana L. - witch-hazel

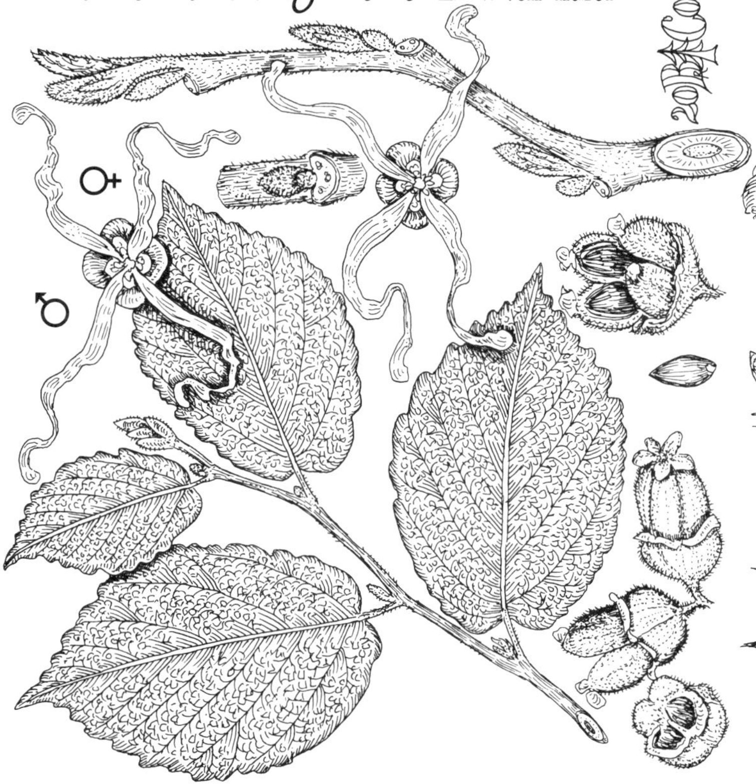

Juglans nigra L. - black walnut

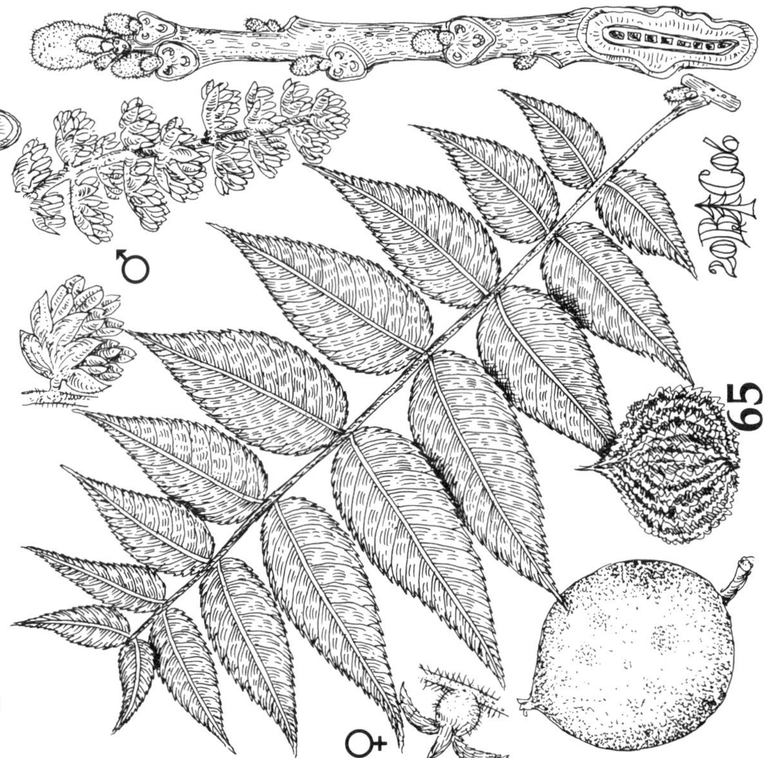

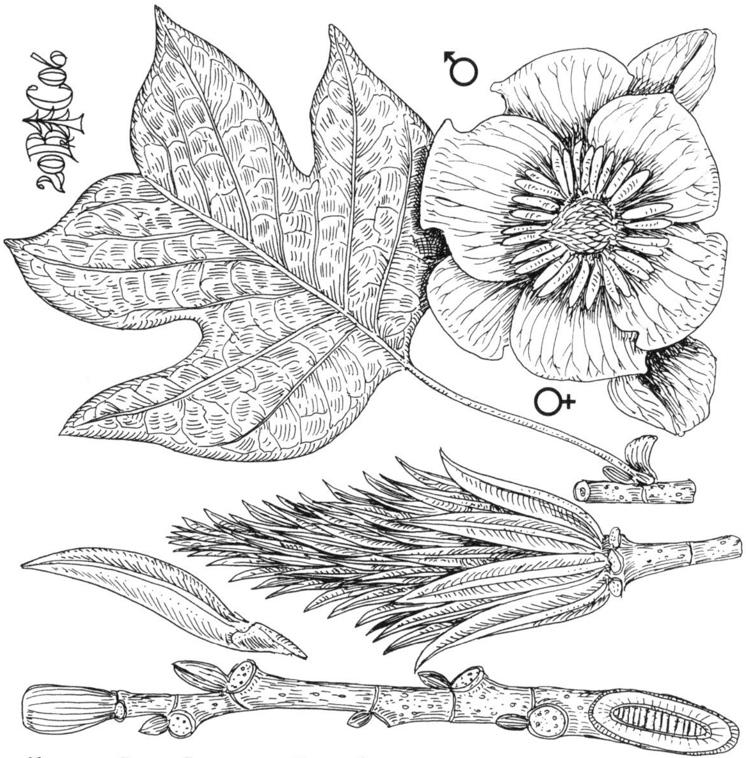

Liriodendron tulipifera L. - yellow-poplar

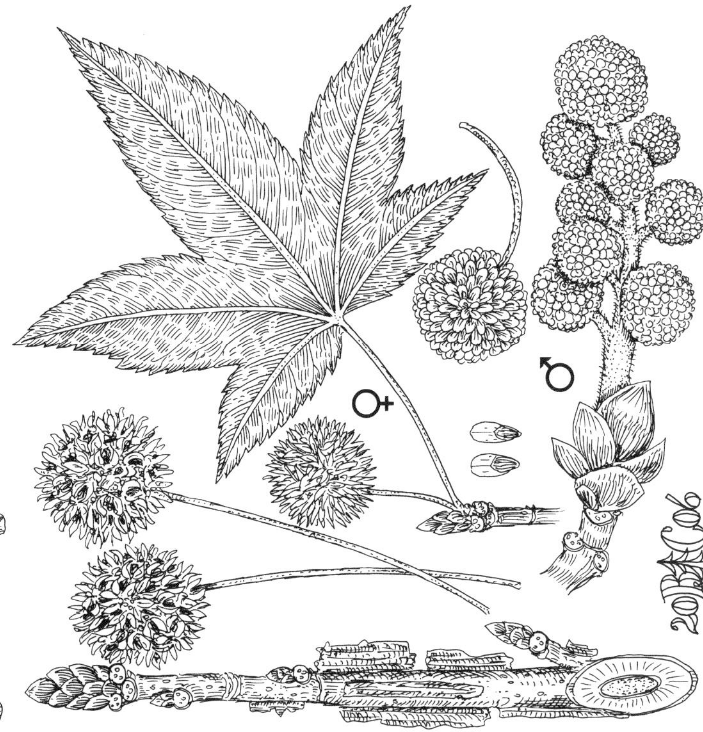

Liquidambar styraciflua L. - sweetgum

Morus rubra L. - red mulberry

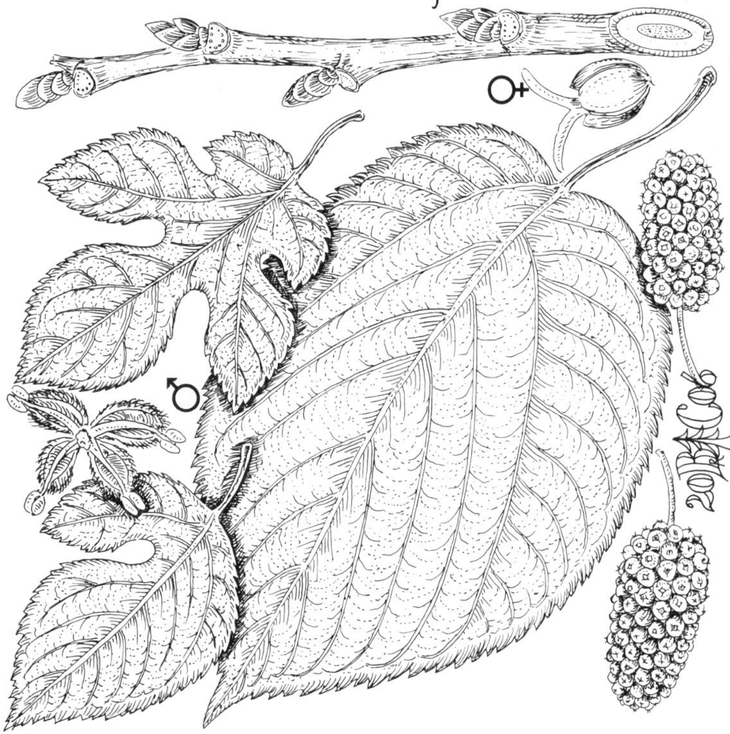

Nyssa sylvatica Marsh. - black tupelo

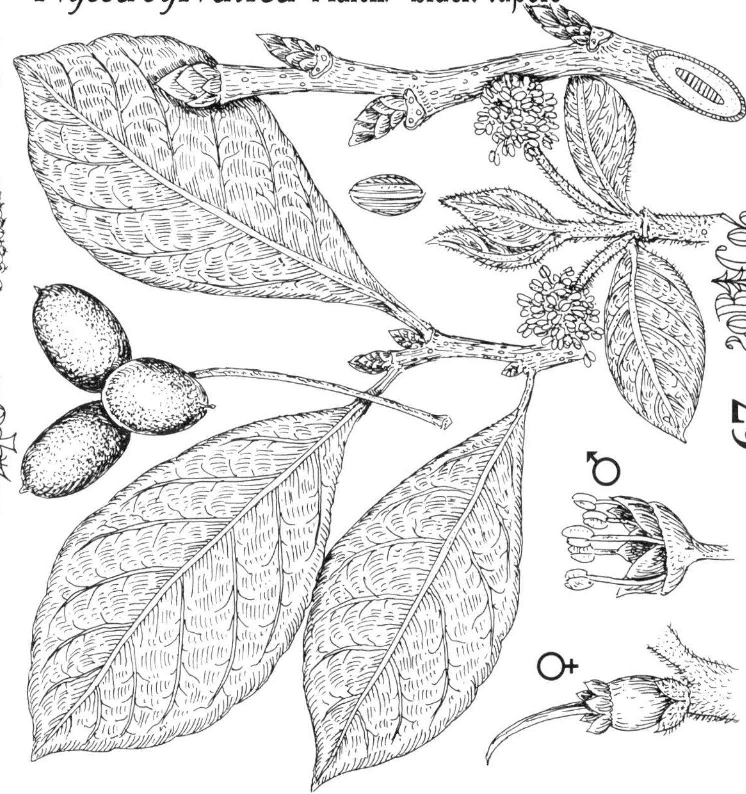

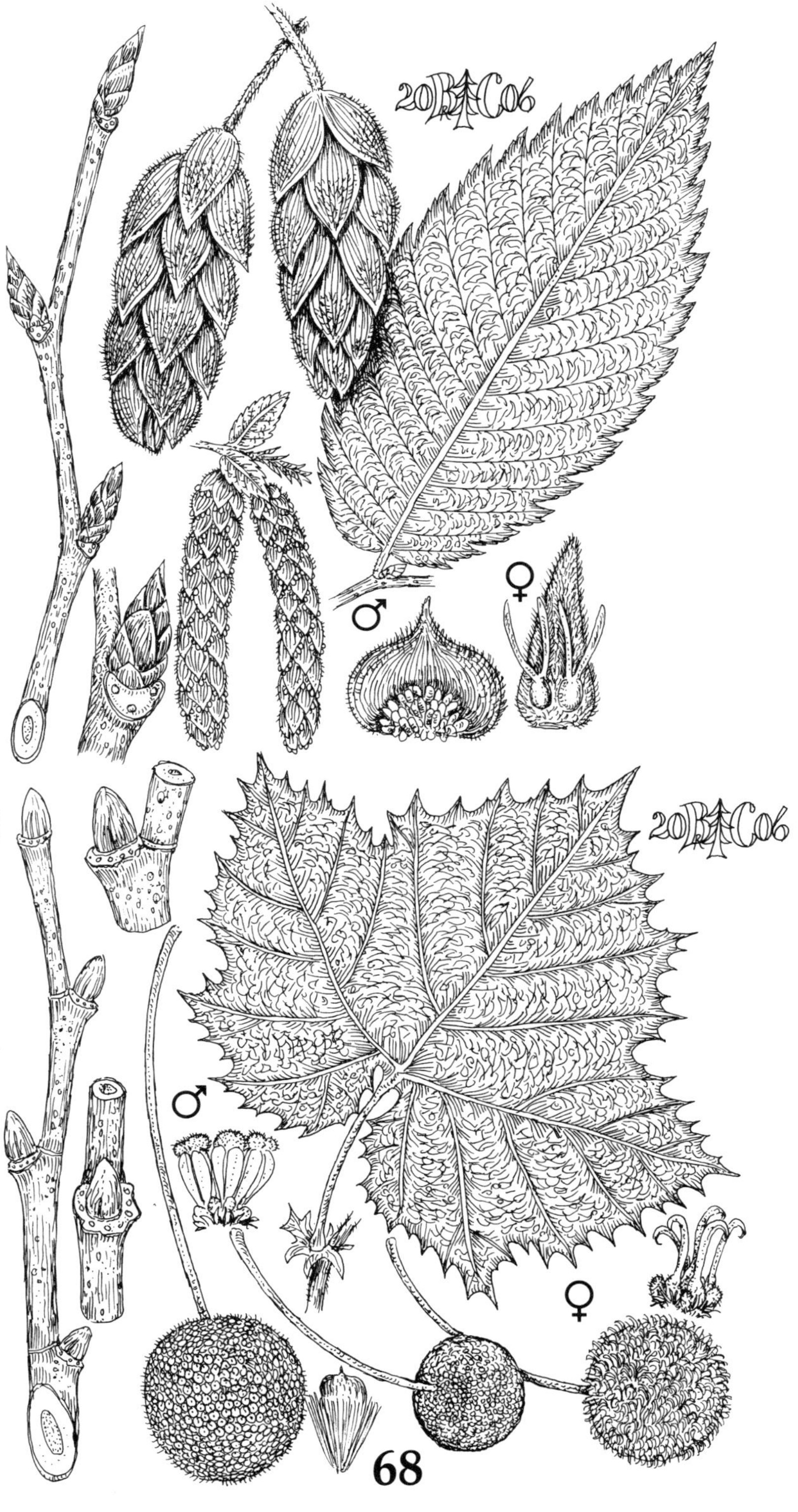

Ostrya virginiana (Mill.) K.Koch - eastern hophornbeam

Platanus occidentalis L. - American sycamore

Populus deltoides Bartr. - eastern cottonwood

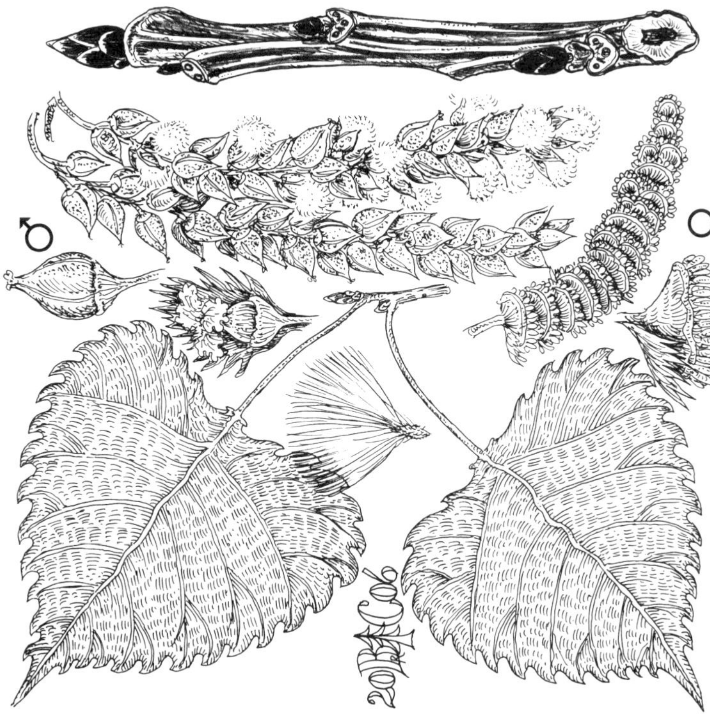

Prunus serotina Ehrh. - black cherry

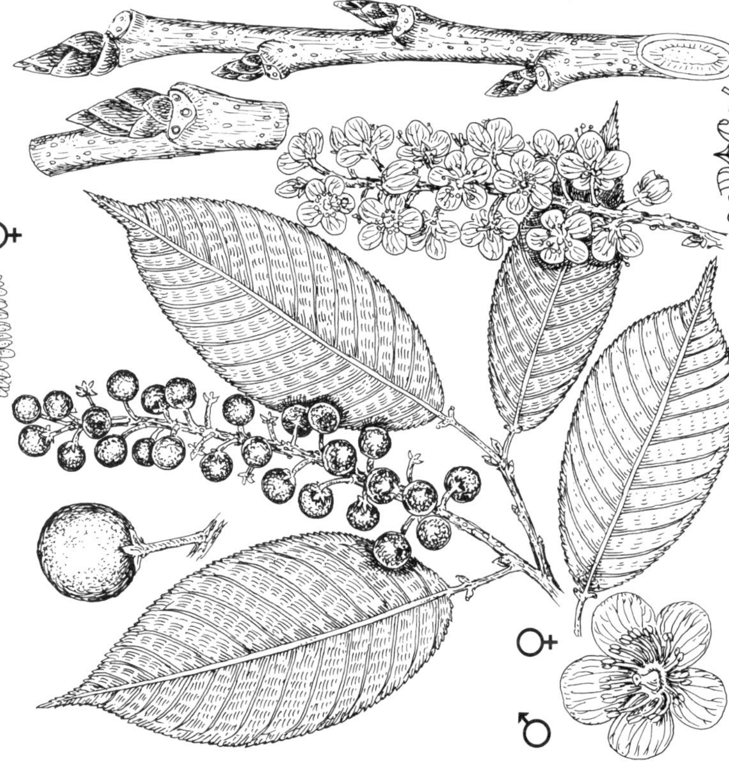

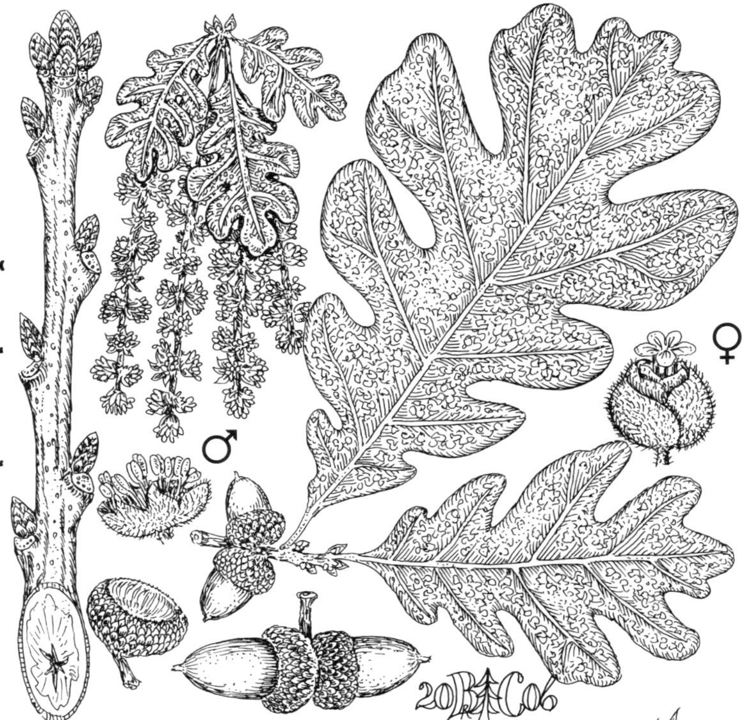

Quercus alba L. - white oak

Quercus muehlenbergii Engelm. - chinkapin oak

Quercus stellata Wangenh. - post oak

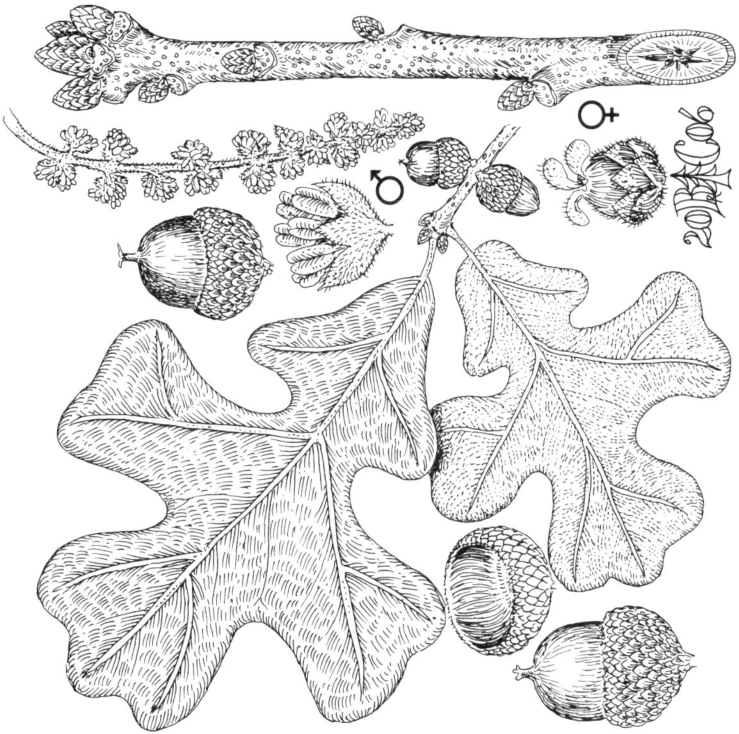

Rhus glabra L. - smooth sumac

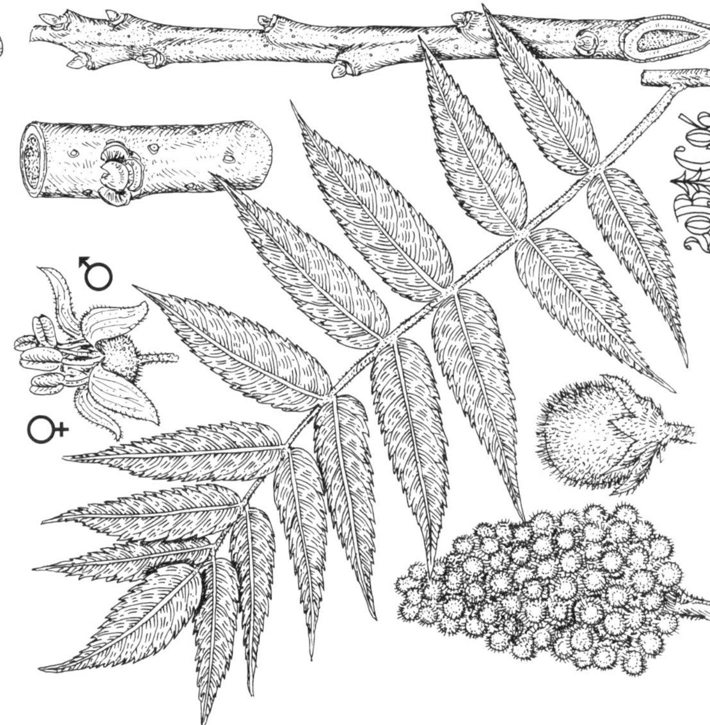

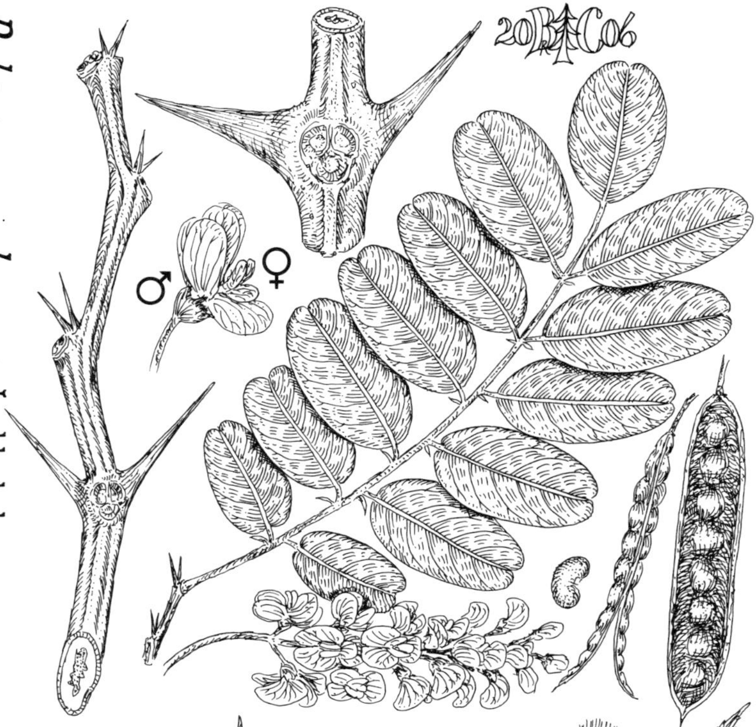

Robinia pseudoacacia L. - black locust

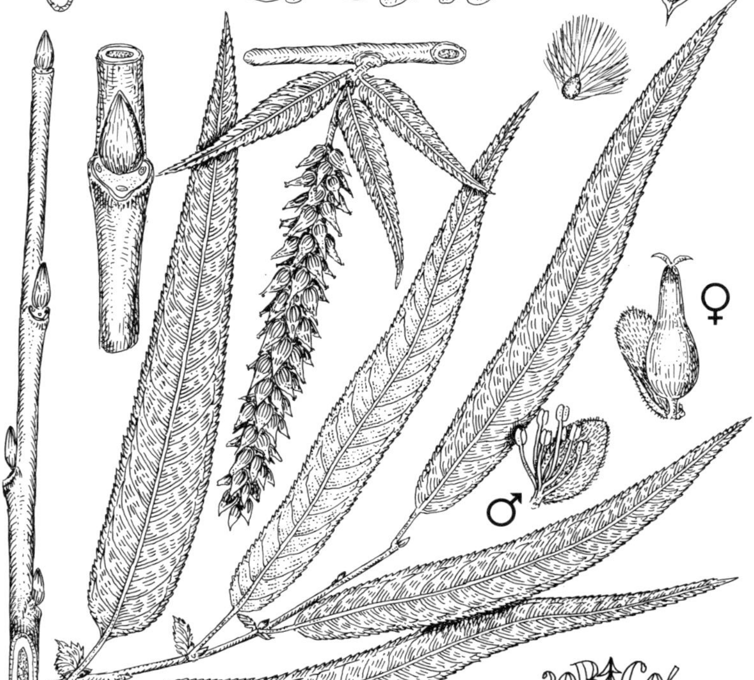

Salix nigra Marsh. - black willow

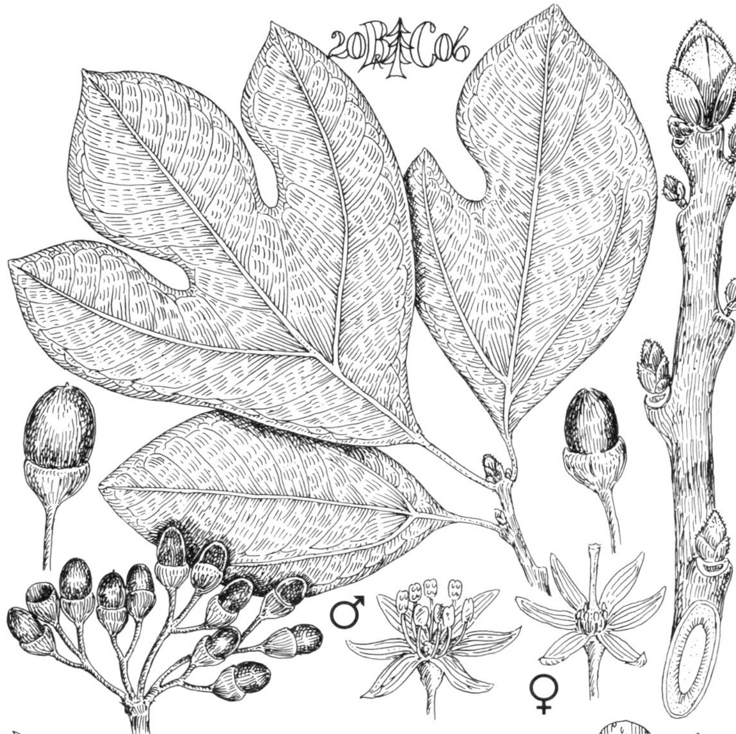

Sassafras albidum (Nutt.) Nees - sassafras

Tilia americana L. - American basswood

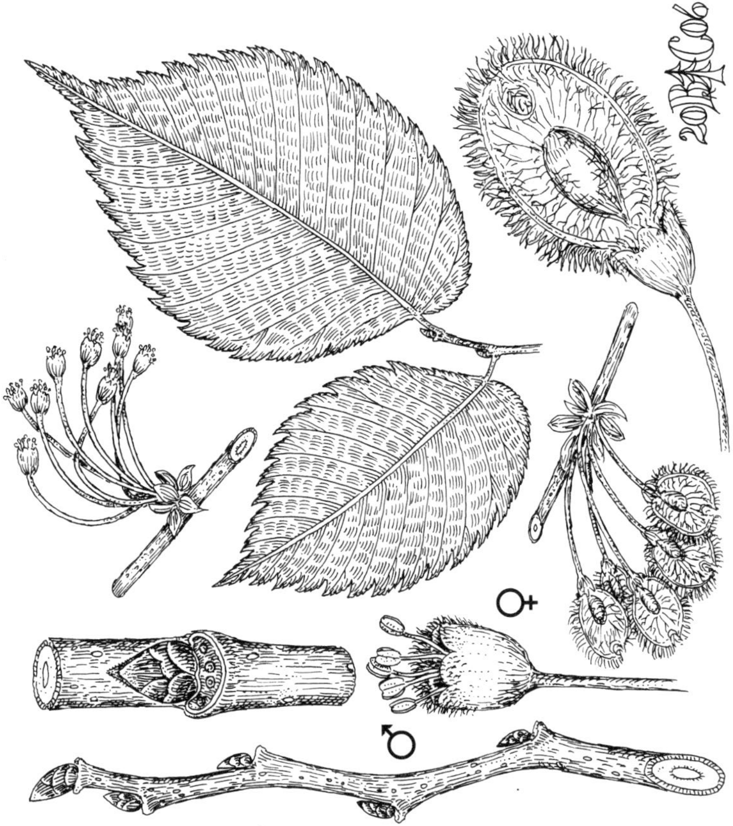

Ulmus americana L. - American elm

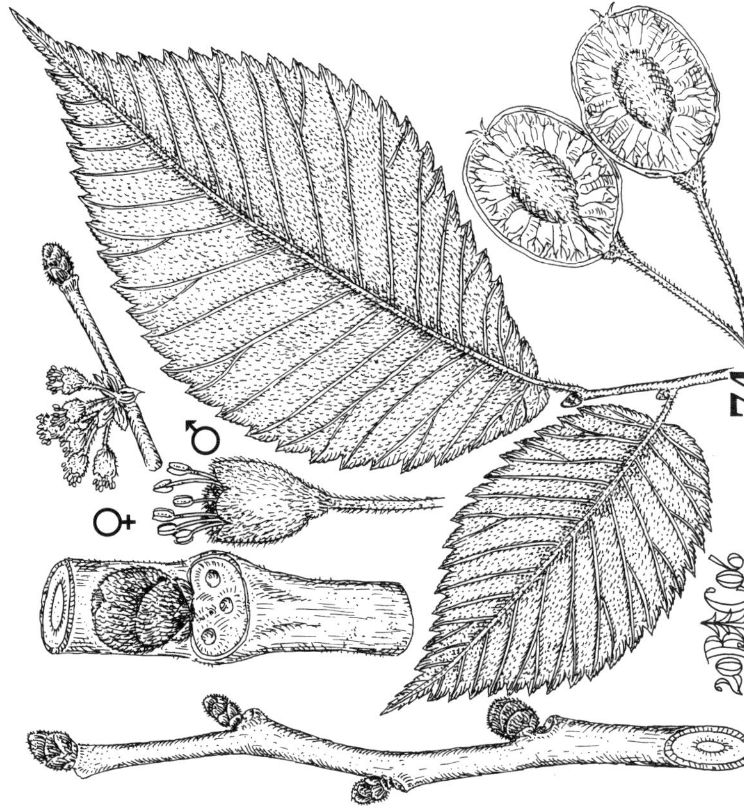

Ulmus rubra Muhl.. - slippery elm

INDEX of VERNACULAR (Common) NAMES

( ) = illustrated page

F

G

H

L

M

N

O

P

R

S

T

U

V

W

Y

INDEX of SCIENTIFIC NAMES

( ) = illustrated page

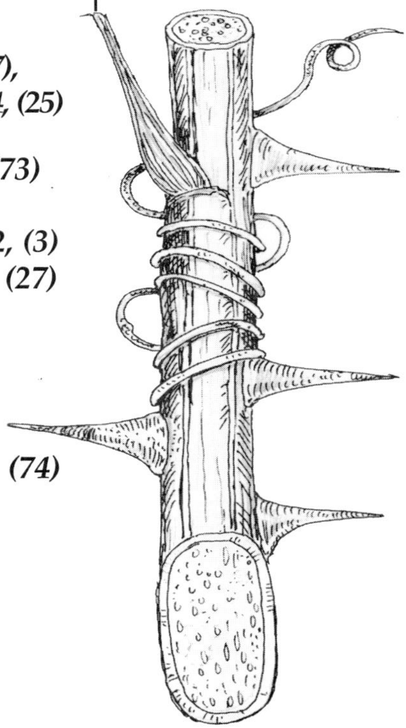

Smilax spp.

POSTER PUBLICATONS

This poster is in full color. Size: 18"by 24"
by Bruce Lyndon Cunningham

BRUCE LYNDON CUNNINGHAM PRODUCTION

Price per Poster - $8.00 or 10 for $49.50
To Order Larger Quantities - Contact:
www.forester-artist.com/936.569.6965

POSTER PUBLICATONS

This poster is in full color. Size: 18"by 24"
by Bruce Lyndon Cunningham

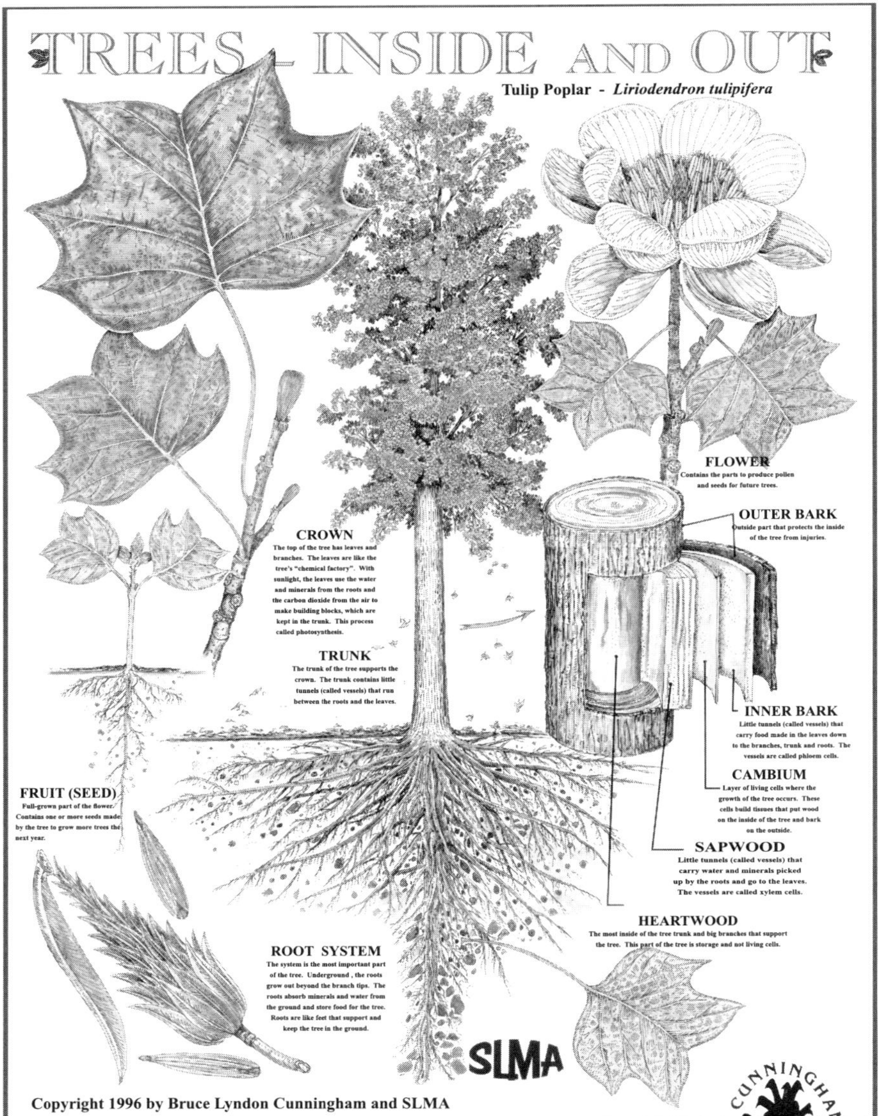

Price per Poster - $8.00 or 10 for $49.50
To Order Larger Quantities - Contact:
www.forester-artist.com/936.569.6965

BRUCE LYNDON CUNNINGHAM PRODUCTION

POSTER PUBLICATONS

This poster is in full color. Size: 18"by 24"
by Bruce Lyndon Cunningham

Price per Poster - $8.00 or 10 for $49.50
T-shirts: $17 or 2 for $32. Size: S, M, L, XL.
Notecards/Envelopes Pkg.: 4 - $8; 10 - $19; 25 - $39.
Contact:www.forester-artist.com/936.569.6965

POSTER PUBLICATONS

This poster is in full color. Size: 18"by 24"
by Bruce Lyndon Cunningham

Price per Poster - $8.00 or 10 for $49.50
To Order Larger Quantities - Contact:
www.forester-artist.com/936.569.6965

BRUCE LYNDON CUNNINGHAM PRODUCTION

BOOK PUBLICATONS

THIS BOOK & COVER ARE IN FULL COLOR.

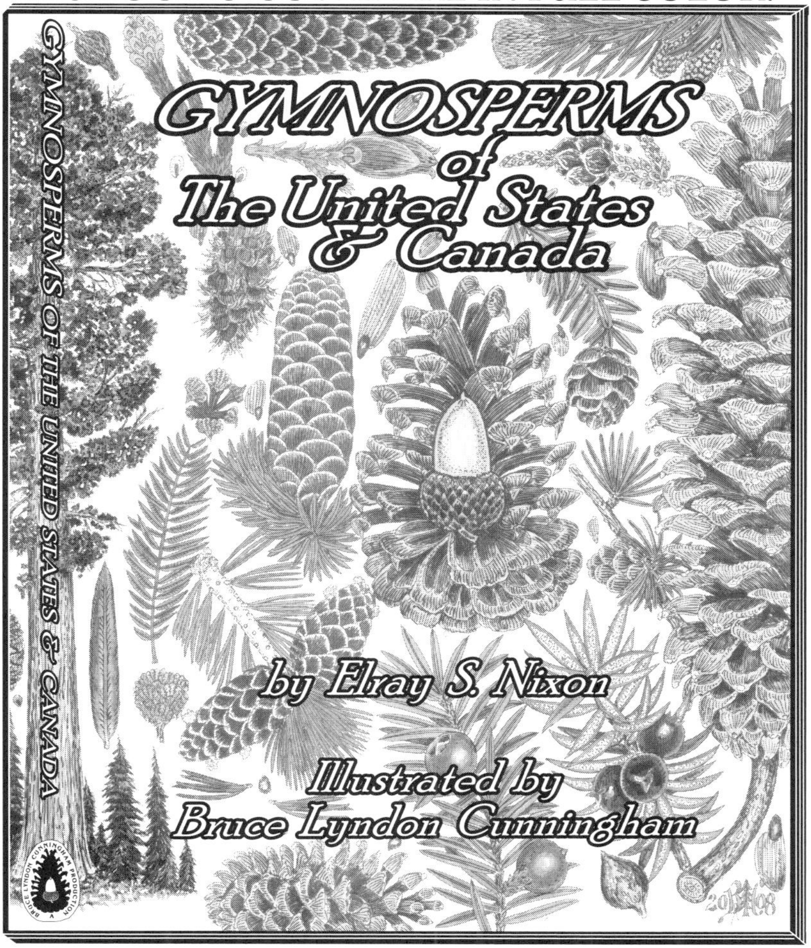

- Contains Identification Keys & Colored Illustrations for 115 gymnosperm species in the United States & Canada.
- Contains an Illustrated Glossary of Plant Structures with over 100 Terms.
- Contains Keys that rely mostly on Cone, Leaf, Bud & Bark Characteristics.
- Can be used by Botanists, Foresters, Naturalists, & Plant Lovers.

Size: 8.5" by 11" * Number of Pages: 208 * 115 Gymnosperm Species in Color

ISBN-13: 978-0-934115-03-2 * Paper Back * $59.95 * LCCN: 2008906099
ISBN-13: 978-0-934115-05-6 * Hard Cover * $69.95 * LCCN: 2008906458

BOOK PUBLICATONS

THIS COVER IS IN FULL COLOR.

- ***Contains Keys & Illustrations for over 300 Woody Species in East Texas.***
- ***Contains an Illustrated Glossary of Plant Structures with over 200 Terms.***
- ***Contains Keys that rely mostly on Leaf, Twig, & Stem Characteristics.***
- ***Can be used by Botanists, Foresters, Naturalists in nearby States to the North and East.***

Size: 8.5" by 11" * Number of Pages: 272 * Over 500 Black/White Illustrations

ISBN-13: 978-0-934115-01-8 * Paper Back * $65.00 * LCCN: 99-91817

BOOK PUBLICATONS

THIS COVER IS IN FULL COLOR.

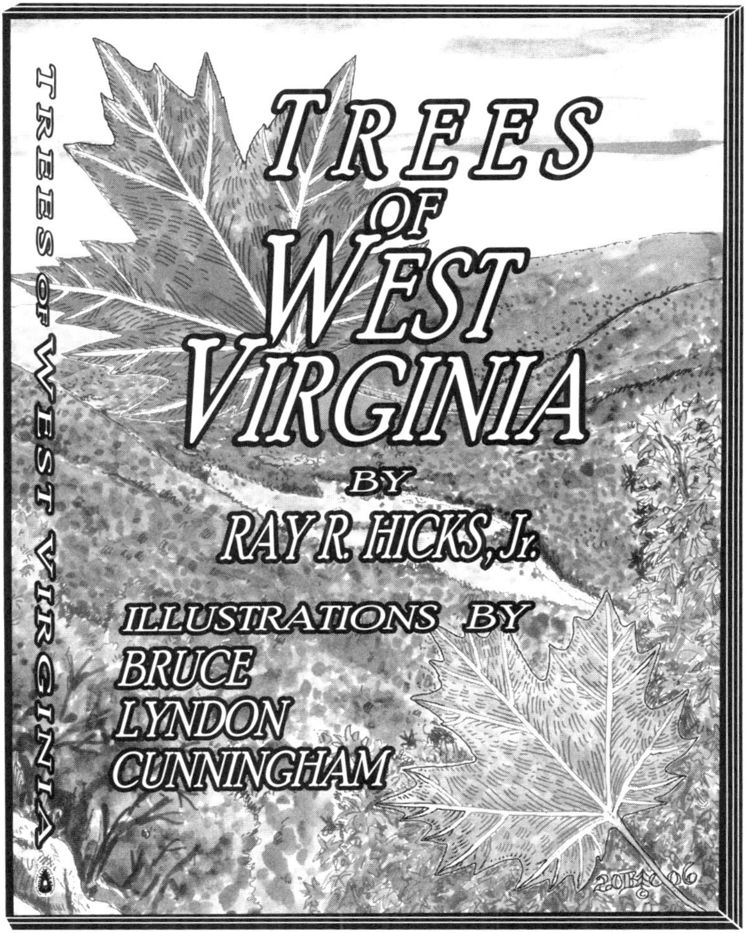

- *Contains a Quick Key & Illustrations for 72 Trees found in West Virginia.*
- *Contains an Glossary of Plant Structures with over 50 Terms with some Illustrations.*
- *Contains a List of the Champion Trees in West Virginia.*
- *Can be used by Botanists, Foresters, Naturalists & Plant Lovers in nearby States.*

Size: 5.5″ by 8.5″ * Number of Pages: 96 * Over 90 Black/White Illustrations
ISBN-13: 978-0-934115-02-5 * Paper Back * $18.95 * LCCN: 2007901142

WOOD CARVINGS

No. 3 Station of the Cross
Wood: *Quercus falcata*
Price: $400 to $950 per station - 16″by 12″

DOGWOOD CROSS

Made from Flowering Dogwood - *Cornus florida*
Size: 6″by 9″
Price: $75

SIGN made from: *Juniperus virginiana, Taxodium distichum , Catalpa bignonioides.*
Size,36″by 38″
Price: $1660+ (One Side)

AMERICAN CHESTNUT TREE FARM
GLEN EYRIE FARM est. 1883

Door Made from:
Juniperus virginiana, and *Pseudotsuga menzieii* - Douglas-fir. Size: 36″by 80″
Price: $1975+ (One Side)

Made from 6 Different Woods - *Quercus falcata, Pinus taeda, Juniperus virginiana, Fraxinus americana, Ilex opaca, and Juglans nigra.* Size: 42″by 42″ - Price: $1850+

Alaskan Sawmill
Quimbey Photo

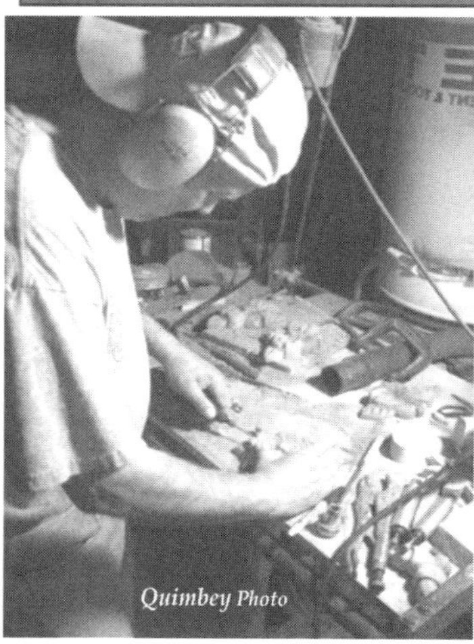

Quimbey Photo

Bear Puppet

Made from - *Fraxinus americana Juglans nigra* Size: 4″x 5″x 3″
Price: $125

936.569.6965
www.forester-artist.com